ORBITAL INTEGRALS ON REDUCTIVE LIE GROUPS AND THEIR ALGEBRAS

Authored by **Francisco Bulnes**

Orbital Integrals on Reductive Lie Groups and Their Algebras
http://dx.doi.org/10.5772/56023
Authored by by Francisco Bulnes

Published by InTech

Janeza Trdine 9, 51000 Rijeka, Croatia

Publishing Process Manager Danijela Duric

Technical Editor InTech DTP Team

Cover InTech Design Team

Additional hard copies can be obtained from orders@intechopen.com

Orbital Integrals on Reductive Lie Groups and Their Algebras, Authored by Francisco Bulnes
p. cm.
ISBN 978-953-51-1007-1

For EU product safety concerns:
IN TECH d.o.o., Prolaz Marije Krucifikse Kozulić 3, 51000 Rijeka, Croatia,
info@intechopen.com or visit our website at intechopen.com.

Meet the author

Francisco Bulnes, doctorate in mathematics sciences by UNAM with doctoral dissertation: "Some Relations between the Vogan-Zuckerman Cohomological Induction and Langlands Classification". Active member of research doctoral seminar "Representations Theory of Real Reductive Lie Groups"(2000-2012). Many works published, badges and recognitions received for this seminar. Mathematics professor in many universities along 20 years. Coordinator of mathematics (2000-2002). Specialist in: Mathematical Physics, Integral Geometry, Field Theory, and Infinite Lie Theory. Principal and scientific advisor in diverse international and national scientific and academic events in the mathematics area. Full member of ANS (American Nano Society). Author of many texts and specialized books all with ISBN of mathematics for scientist and engineers, and published for many universities inside and without from Mexico. Principal of Appliedmath (2005-2008). Author of diverse international and national referred Journal articles in many areas of pure and applied mathematics (http://www.scirp.org/journal/PaperInformation.aspx?paperID=20338, http://www.scirp.org/journal/apm/) for example. Quantum Electrodynamics expert by the (AIP) American Institute of Physics. Co-Principal of doctoral seminar of Lie groups in IM/UNAM and fellowships and other distinctions in the doctoral Seminar of research. International projects judge of mathematics to CONACYT/Mexico. Book-Editor for other mathematicians. Scientific consul in research on mathematics and advanced systems with some countries as Cuba, USA and Russia. Actually the Dr. Bulnes is researching on nanotechnology, on the classification problem of irreducible unitary representations and on one new theory of the Universe with MIT and Harvard groups. Editor-in-Chief of Journal on Photonics and Spintronics, of USA, http://www.researchpub.org/journal/jps/editorial%20board.html. Has been involved in several editorial advisory boards of journals in mathematics. Actually is Principal of Research TESCHA from 2009. His interest areas are: mathematical physics, cohomology of cycles (motivic and relative), integral geometry, K-theory, moduli spaces and nanotechnology.

Contents

To my master
Felix Recillas Juarez (1918-2010),
(PhD, Princeton, New Jersey, USA, 1948)
with all gratitude and my
recognition for their wisdom
and friendship

Preface

The following work is the result of diverse explains, courses and conferences dictated by the *Francisco Bulnes, PhD,* in many international academic places and diverse national forums along thirteen years inside the research in integral theories of generalized measures, integral operators cohomology and Integration on Haar spaces and on spaces of infinite dimension. Many of these talks, were given in Institute of Mathematics of National Autonomous University of Mexico (IM/UNAM), into the seminar of "*Representation Theory of Reductive Lie Groups*" (1999-2006) and after, in the Department of Research in Mathematics and Engineering of Technological Institute of High Studies of Chalco (DIMI/TESCHA) among the years 2004-2012.

The purpose is to present a complete course on these topics and to establish some orbital applications of the integral on Lie groups to the harmonic analysis and induced representations in representation theory.

Some other applications in this respect they will be given for some integral models of the cosmos.

Department of Research in Mathematics and Engineering, TESCHA
Institute of Mathematics UNAM, Mexico

Chapter I

Introduction

I. 1. Introduction

In the study of the theory of *irreducible unitary representations*, is necessary to analyze and demonstrate diverse results on integral orbital of functions belonging to the cohomology $H^i(\mathfrak{g}, K; V \otimes V_\gamma^*)$, and that it is wanted they belong to the *$L^2(G)$-cohomology* of their reducible unitary representations called *discrete series*. Then is necessary consider the *Frèchet space* I(G), and analyze the *2-integrability* to the fibers of the space G/K, in spaces or locally compact components of G/K. For it will be useful the invariance of the corresponding measures of Haar under the actions of Ad(G), and the corresponding images of the *Harish-Chandra transform* on the space of functions $I_{a,b}(G)$.

Likewise, we will obtain a space in *cuspidal forms* that is an introspection of the class of the discrete series in the whole space G.

This *harmonic analysis* in the context of the space in cuspidal forms is useful in the exploration of the behavior of *characters* for those($\mathfrak{g}$, K)-*modules* $H^i(\mathfrak{g}, K; V \otimes V_\gamma^*)$ and also for the generalization of the *integral formula of Plancherel* on *locally compact spaces* of G.

The generalization of the Plancherel formula is useful for the study of the functions on symmetrical spaces.

I. 2. Generalized spheres on Lie groups

We consider to $G = L/H_g$, a *homogeneous space* with origin $o = \{H_g\}$. Given $g_o \in G$, let L_{go} be the subgroup of G, letting g_o fix, that is to say; the subgroup of isotropy of G, in g_o.

Def. I. 2.1. A *generalized sphere* is an *orbit* $L_{go}g$, in G, of some point $g \in G$, under the subgroup of isotropy in some point $g_o \in G$.

In the case of a Lie group the generalized spheres are the left translations (or right) of their *conjugated classes*.

We assume that H_g, and each L_{go}, is *unimodular*. But is considering $L_{go}g = L_{go}/(L_{go})_g$, such that $(L_{go})_g$, be unimodular then the orbit $L_{go}g$, have an invariant measure determined except for a constant factor. Then are our interest the following general problems:

a. To determine a function f, on G, in terms of their orbital integrals on generalized spheres.

In this problem the essential part consist in the normalization of invariant measures on different orbits.

If is the case in that H_g, is compact, the problem A), is trivial, since each orbit $L_{go}g$, have finite invariant measure such that $f(g_o)$ is given as the limit when $g \rightarrow g_o$, of the variation of f, on $L_{go}g$.

I. 2.1. Orbits

Suppose that to every $g_o \in G$, exist an open set L_{g0}-invariant $C_{go} \subset G$, containing g_o in their classes such that to each $g \in C_{g0}$, the group of isotropy $(L_{go})_g$, is compact. The invariant measure on the orbit $L_{g0}g$ $(g_o \in G,\ g \in C_{go})$ can be normalized consistently as follows: We fix a Haar measure dg_o, on L_o ($H_g = L_o$). If $g_o = g \mid o$, we have $L_{g0} = gL_og^{-1}$,and we can to carry on dg_o, to the measure dg_{g0},on L_{go} through of the conjugation $z \rightarrow gzg^{-1}$ $(z \in L_o)$. Since dg_o, is bi-invariant, $dg_{go,g}$, is independent of the election of g satisfying $g_o = g \mid o$, the which is bi-invariant. Since $(L_{go})_g$, is compact, this have an only measure of Haar $dg_{go,g}$, with total measure 1 and reason why dg_o, and $dg_{go,g}$, determine canonically an invariant measure μ on the orbit$L_{g0}g = L_{go}/(L_{go})_g$.

Reason why also the following problem can to establish:

b. To express to $f(g_o)$, in terms of the integrals$\int_{L_{g0}g} f(p)d\mu(p)$, $g \in C_{go}$.

that is to say, the calculus of the orbital integrals on those measurable open sets called orbits.

I. 3. Invariant measures on homogeneous spaces

Let G, a locally compact topological group. Then a left invariant measure on G, is a positive measure, dg, on G, such that

$$\int_G f(xg)dg = \int_G f(g)\, dg, \tag{I. 3.1}$$

$\forall x \in G$, and all $f \in C_c(G)$. If G, is separable then is acquaintance (Haar theorem) that such measure exist and is unique except a multiplicative constant.

If G, is a Lie group with a finite number of components then a left invariant measure on G, can be identified with a left invariant n-form on G (where dim $G = n$). If μ, is a left invariant non-vanishing n-form on G, then the identification is implemented by the integration with regard to μ, using the canonical method of differential geometry. If G, is compact then we can (if is not specified) to use normalized left measures. This is those whose measure total is 1.

If dg is a left invariant measure and if $x \in G$, then we can define a new left invariant measure on G, μ_x, as follows:

$$\mu_x(f) = \int_G f(xg)dg, \tag{I. 3.2}$$

The uniquely of the left invariant measure implies that

$$\mu_x(f) = \delta(x)\int_G f(xg)dg, \tag{I. 3.3}$$

with δ, a function of x, which is usually called the modular function of G. If δ, is identically equal to 1,then we say that G, is unimodular. If G is then unimodular we can call to a left invariant measure (which is automatically right invariant) invariant. It is not difficult affirm that δ, is a continuous homomorphism of G, in the multiplicative group of positive real numbers. This implies that if G, is compact then G, is unimodular.

If G, is a Lie group, the modular function of G, is given by the following formula:

$$\delta(x) = |\det \mathrm{Ad}(x)|, \tag{I. 3.4}$$

where Ad, is the usual adjunct action of G, on their Lie algebra.

Let M, be a soft manifold and be μ, their form of volume. Let G, be a Lie group acting on M. Then $(g^*\mu)_x = c(g, x)\mu_x$, each $g \in G$, and $x \in M$. If is left as exercise verify that c satisfies the cocycle relationship

$$c(gh, x) = c(g, hx)c(h, x) \quad \forall\ h, g \in G, x \in M, \tag{I. 3.5}$$

We write as $\int_M f(x)dx$, to $\int_M f\mu$. The usual formula of change of variables implies that

$$\int_M f(gx)|c(g, x)|dx = \int_G f(x)dx, \tag{I. 3.6}$$

to $f \in C_c(G)$, and $g \in G$.

Let H, be a closed subgroup of G. Be M = G/H. We assume that G, have a finite number of connect components. A G-invariant measure, dx, on M is a measure such that

$$\int_M f(gx)dx = \int_M f(x)dx, \quad \forall\ f \in C_c(G), g \in G \tag{I. 3.7}$$

If dx, comes of a form of volume on M, then (I. 3.7), is the same, which is equal to that $|c(g, x)| = 1\ \forall\ g \in G, x \in M$.

If M, is a soft manifold then is well acquaintance that M, have a form of volume or M, have a double covering that admit a form of volume. To rising of functions to the double covering (if it was necessary) one can integrate relatively to a form of volume on any manifold. Come back to the situation M = G/H, is not difficult demonstrate that M, admit a measure G-invariant if and only if the unimodular function of G, restricted to H, is equal to the unimodular function of H. Under this condition, a measure G-invariant on M is constructed as follow: Be $\mathfrak{g}$, the Lie algebra of G, and be $\mathfrak{h}$, the subalgebra of $\mathfrak{g}$, corresponding to H. Then we can to identify the tangent space of $1H$ to M with $\mathfrak{g}/\mathfrak{h}$. The adjunct action of H, on $\mathfrak{g}$, induces to action $Ad^{\sim}$, of H, on $\mathfrak{g}/\mathfrak{h}$. The condition mentioned to obtain an identity in (I. 3. 7) tell us that $|Ad^{\sim}(h)| = 1\ \forall h\in H$. Thus if H^0, is the identity component of H, (as is usual) and if μ, is a element not vanishing of $\Lambda^m(\mathfrak{g}/\mathfrak{h})^*$[3] (m = dim G/H) it is can to translate μ, to a form of G-invariant volume on G/H^0.

Therefore for rising of functions of M, to G/H^0, is had an invariant measure on M. But the Fubini theorem affirms that we can normalize dg, dh and dx, such that

$$\int_G f(g)dg = \int_{G/H}\left(\int_H f(gh)\, dh\right) d(gH), \quad f\in C_c(G) \tag{I. 3.8}$$

Let G, be a Lie group with a finite number of connects components. Let H, be a closed subgroup of G, and let dh, be a selection of left invariant measure on H. The following result is used in the calculus of measures on homogeneous spaces.

Lemma I. 2.1. If f, is a compactly supported continuous function on H/G, (note the change to the right classes) then it exists, g, a continuous function supported compactly on such G, that

$$f(Hx) = \int_G g(hx)\, dh, \tag{I. 3.9}$$

This result is usually demonstrated using a "partition of the unit" as principal argument.

For details of demonstration see [1].

Let G, be a Lie group and be A, and B, subgroups in G, such that A, and B, are compact and such that G = AB. The following result is used to the study of induced representations and classes of induced cohomology.

Lemma I. 2.2. We asume that G, is unimodular. If da, is a left invariant measure on A, and db, is a left invariant measure on B, then we can elect an invariant measure, dg, on G, such that

$$\int_G f(g)\, dg = \int_{A\times B} f(ab)\, da\, db, \quad \text{para } f\in C_c(G) \tag{I. 3.10}$$

Proof: Consult [2]. ■

In the following section we will explain basic questions on invariant measures on homogeneous spaces. With it will stay clear the concept and use of normalized measures.

Let G, be a Lie group with Lie algebra $\mathfrak{g}$; let H, a closed subgroup with Lie algebra $\mathfrak{h}\subset\mathfrak{g}$. Each $x\in G$, gives rise to an analytic diffeomorphism

$$\tau(x): gH\to xgH, \tag{I. 3.11}$$

of G/H, onto itself. Let π, denote the natural mapping of G, onto G/H, and put $o = \pi(e)$. If $h\in H$, $(d\tau(h))_o$, is an endomorphism of the tangent space $(G/H)_o$. For simplicity, we shall write $d\tau(h)$, instead of $(d\tau(h))_o$, and $d\pi$, instead of $(d\pi)_e$ [4].

Lemma I. 2.3.

$$\det(d\tau(h)) = \frac{\det Ad_G(h)}{\det Ad_H(h)}, \tag{I. 3.12}$$

$\forall h\in H$.

Proof. $d\pi$, is a linear mapping of $\mathfrak{g}$, onto $(G/H)_o$, and has kernel $\mathfrak{h}$. Let $\mathfrak{m}$, be any subspace of $\mathfrak{g}$, such that $\mathfrak{g} = \mathfrak{h} + \mathfrak{m}$, (direct sum). Then $d\pi$, induces an isomorphism of $\mathfrak{m}$, onto $(G/H)_o$. Let $X\in\mathfrak{m}$. Then $Ad_G(h)X = dR_{h-1}\circ dL_h(X)$. Since $\pi\circ R_h = \pi$, $\forall\ h\in H$, and $\pi\circ L_g = \tau(g)\circ\pi$, $\forall\ g\in G$, we obtain

$$d\pi \circ Ad_G(h)X = d\tau(h) \circ d\pi(X), \tag{I. 3.13}$$

The vector $Ad_G(h)X$, decomposes according to $\mathfrak{g} = \mathfrak{h} + \mathfrak{m}$,

$$Ad_G(h)X = X(h)_{\mathfrak{h}} + X(h)_{\mathfrak{m}}, \tag{I. 3.14}$$

The endomorphism

$$A_h : X \to X(h)_{\mathfrak{m}}, \tag{I. 3.15}$$

of $\mathfrak{m}$, satisfies

$$d\pi \circ Ad_h(X) = d\tau(h) \circ d\pi(X), \tag{I. 3.16}$$

$\forall\ X\in\mathfrak{m}$, so $\det A_h = \det(d\tau(h))$. For other side,

$$exp\ Ad_G(h)tT = h\ exp\ tTh^{-1} = \exp exp\ Ad_H(h)tT, \tag{I. 3.17}$$

for $t\in\mathbb{R}$, $T\in\mathfrak{h}$. Hence $Ad_G(h)T = Ad_H(h)T$, so

$$\det Ad_G(h) = \det A_h \det Ad_H(h), \tag{I. 3.18}$$

and the lemma is proved. ■

Proposition I. 2.1. Let m = dim G/H. The following conditions are equivalent:

i. G/H, has a nonzero G-invariant m-form ω;
ii. $\det Ad_G(h) = \det Ad_H(h)$, for $h\in H$.

If these conditions are satisfied, then G/H, has a G-invariant orientation and the G-invariant m-form ω, is unique up to a constant factor.

Proof. Let ω, be a G-invariant m-form on G/H, $\omega \neq 0$. Then the relation $\tau(h)^*\omega = \omega$, [3] at the point o, implies $\det(d\tau(h)) = 1$, so ii), holds. For other side, let $X_1, \ldots, X_m$, be a basis of $(G/H)_o$, and let $\omega^1, \ldots, \omega^m$, be the linear functions on $(G/H)_o$, determined by $\omega^i(X_j) = \delta_{ij}$. Consider the element$\omega^1\wedge\ldots\wedge\omega^m$, in the Grassmann algebra of the tangent space $(G/H)_o$. The condition ii), implies that $\det(d\tau(h)) = 1$, and the element $\omega^1\wedge\ldots\wedge\omega^m$, is invariant under the linear transformation $d\tau(h)$. It follows that exists a unique G-invariant m-form ω, on G/H, such that $\omega_o = \omega^1\wedge\ldots\wedge\omega^m$. If ω^*, is another G-invariant m-form on G/H, then $\omega^* = f\omega$, where $f\in C^\infty(G/H)$. Owing to the G-invariance, f = *constant*.

Assuming i), let $\varphi : p \to (x_1(p), \ldots, x_m(p))$, be a system of coordinates on an open connected neighborhood U, of $o\in G/H$, on which ω, has an expression

$$\omega_U = F(x_1,\ldots,x_m)\, dx_1\wedge\ldots\wedge dx_m,$$

With $F > 0$, The pair $(\tau(g)U, \varphi \circ \tau(g^{-1}))$, is a local chart on a connected neighborhood of $g \bullet o\in G/H$. We put $(\varphi \circ \tau(g^{-1}))(p) = (y_1(p), \ldots, y_m(p))$, for $p\in\tau(g)U$. Then the mapping

$$\tau(g) : U \to \tau(g)U,$$

has expression

$$(y_1, \ldots, y_m) = (x_1, \ldots, x_m).$$

On $\tau(g)U$, ω, has an expression

$$\omega_{\tau(g)U} = G(y_1,\ldots, y_m)\, dy_1\wedge\ldots\wedge dy_m,$$

and since $\omega_q = \tau(g)^*\omega_{\tau(g)q}$, we have for $q\in U \cap\tau(g)U$,

$$\omega_q = G(y_1(q),\ldots, y_m(q))\, (dy_1\wedge\ldots\wedge dy_m)_q = G(x_1(q), \ldots, x_m(q))\, (dx_1\wedge\ldots\wedge dx_m)_q,$$

Hence $F(x_1(q), \ldots, x_m(q)) = G(x_1(q), \ldots, x_m(q))$, and

$$F(x_1(q), \ldots, x_m(q)) = F(y_1(q),\ldots, y_m(q))[\partial(y_1(q),\ldots, y_m(q))/\partial(x_1(q), \ldots, x_m(q))],$$

which shows that the Jacobian of the mapping $(\varphi \circ \tau(g^{-1})) \circ \varphi^{-1}$, is positive. Consequently, the collection $(\tau(g)U, \varphi \circ \tau(g^{-1}))_{g\in G}$, of local charts turns G/H, into an oriented manifold and each $\tau(g)$, is orientation preserving. Then G-invariant form ω, now gives rise to an integral $\int f\omega$, which is invariant in the sense that

$$\int_{G/H} f\omega = \int_{G/H} (f \circ \tau(g))\omega, \quad \forall\ g\in G.$$

However, just as the Riemannian measure did not require orientability; an invariant measure can be construct on G/H, under a condition which is slightly more general than (ii). The projective$\mathbb{P}^2$, will, for example, satisfy this condition whereas it does not satisfy (ii). We recall that a measure μ, on G/H, is said to be invariant (or more precisely *G*-invariant) if $\mu(f \circ \tau(g)) = \mu(f)$, for all $g\in G$. ■

Theorem I. 2.1. Let *G*, be a Lie group and H, a closed subgroup. The following relation is satisfied

$$|det\ Ad_G(h)| = |det\ Ad_H(h)|, \qquad h \in H, \tag{I. 3.19}$$

Is a necessary and sufficient condition for the existence of a G-invariant positive measure on G/H. This measure dg_H, is unique (up to a constant factor) and

$$\int_G f(g)dg = \int_{G/H}\left(\int_H f(gh)dh\right)dg_H, \quad \forall\ f \in C_c(G), \tag{I. 3.20}$$

if the left invariant measures dg, and dh, are suitably normalized.

Proof [6], [1]

Chapter II

Integrals, Functional and Special Functions on Lie Groups and Lie Algebras

II. 1. Spherical functions

Let $P = P_{\Phi}$, be the minimal parabolic subgroup of G, with the Langlands decomposition

$$P = {}^{0}MAN, \tag{II. 1.1}$$

If (σ, H^{σ}), is an irreducible unitary representation of ${}^{0}M$, and if $\mu\in(\mathfrak{a}_{C})^{*}$, then $(\pi_{\sigma,\mu}, H^{\sigma,\mu})$, can denote the corresponding representation in principal serie. $H^{\sigma,\mu}$, is equivalent with $I_K(\sigma) = H^{\sigma}$. Indeed, if $H^{\sigma,\mu}$, is a representation of K, then exist $\sigma\in G$, such that $\sigma\mathfrak{a}^{*}= \mathfrak{a}^{*}$, to $\mathfrak{a}^{*}$, the dual algebra of the algebra $\mathfrak{a} \subset \mathfrak{p}$, since always there is a maximal Abelian subalgebra in $\mathfrak{p}$. Then $I_K(\sigma\mu) = I_K(\mu) = H^{\sigma,\mu}$. By the subcocient theorem to induced representations using the Casselman theorem, is possible to construct an operator that go from $Hom_{\mathfrak{g},K}(V, H^{\sigma})$, to $Hom_{\mathfrak{g},K}(V, H^{\sigma,\mu})$, that define an unitary equivalence between the representations in H^{σ}, and $H^{\sigma,\mu}$. Then H^{σ}, and $H^{\sigma,\mu}$, are equivalent representations as representations of the group K.

If $f\in H^{\sigma}$, that is to say, $f\in L^2({}^{0}M/K)$, then $f_{\mu}(nak) = a^{\mu+\rho}f(k)$, $\forall$ $n\in N$, $a\in A$, and $k\in K$.

If $g\in G$, and $g = nak$, with $n\in N$, $a\in A$, and $k\in K$, then we can write $n(g) = n$, $a(g) = a$, $k(g) = k$. The theory of real reductive groups implies that as functions on G, n, a, and k, are smooth functions. We denote as **1**, to the function on K, that is identically equal to 1.

Let γ_0, be the class of the trivial representations of K. Then is clear that

$$(H^{\mu})_K(\gamma_0) = \mathbb{C}I, \tag{II. 1.2}$$

If $\mu\in(\mathfrak{a}_{C})^{*}$, then we define Ξ_{μ}, for

$$\Xi_{\mu}(g) = \langle\pi_{\mu}(g)\mathbf{1}_{\mu}, \mathbf{1}_{\mu}\rangle, \tag{II. 1.3}$$

Said extended function to all the subgroup K, come given as

$$\Xi_{\mu}(g) = \int_K a(kg)^{\mu+\rho}dk, \ \ \forall\ g\in G \tag{II. 1.4}$$

where $\mathbf{1}_{\mu}(g) = a(g)^{\mu+\rho}$, and $\mathbf{1}_{\mu}(k) = a(kg)^{\mu+\rho}$, $\forall$ $g\in G$, $k\in K$.

Proposition II. 1.1. If $s\in W(\mathfrak{g}, \mathfrak{a})$, then $\Xi_{s\mu} = \Xi_{\mu}$, $\forall\mu\in(\mathfrak{a}_{C})^{*}$.

Proof. The demonstration of this result due to Harish-Chandra can require of some previous considerations. Let $C_c(K\backslash G/K)$, the space of all differentiable K-bi-invariants functions on G, of compact support on G. To demonstrate the proposition is enough demonstrate that if $f\in C_c(K\backslash G/K)$, then

$$\int_G f(g)\Xi_\mu(g)\,dg = \int_G f(g)\Xi_{s\mu}(g)\,dg, \qquad \text{(II. 1.5)}$$

Calculating

$$\int_G f(g)\Xi_\mu(g)\,dg = \int_{G\times K} f(g)a(kg)^{\mu+\rho}dgdk = \int_G f(g)\mathbf{1}_\mu(g)\,dg,$$

where f, is invariant left. Let S = An, and let ds, be a election of the left invariant measure on S, applying the measurability in real reductive groups. Then

$$\int_G f(g)\mathbf{1}_\mu(g)\,dg = \int_{S\times K} f(sk)a(s)^{\mu+\rho}dsdk = \int_S f(s)a(s)^{\mu+\rho}ds,$$

due to the K-invariance of f. Now ds, can be normalized such that $ds = a^{-2\rho}dnda$. Thus, calculating said integral, this can to take the form

$$\int_{N\times A} f(na)a^{\mu-\rho}dnda,$$

We call to $F_f(a) = a^{-\rho}\int_N f(na)dn$. Then we have demonstrated that

$$\int_G f(g)\Xi_\mu(g)\,dg = \int_A F_f(a)a^{\mu}da, \qquad \text{(II. 1.6)}$$

Therefore to demonstrate the proposition is enough demonstrate that

$$F_f(\exp H) = F_f(\exp sH),\ \forall\ H\in\mathfrak{a},\ f\in C_c^{\infty}(K\backslash G/K),\ \text{and } s\in W(\mathfrak{g},\mathfrak{a}), \qquad \text{(II. 1. 7)}$$

Indeed, let Φ^+, be a positive roots space in $\Phi(\mathfrak{g}, \mathfrak{a})$, corresponding to $\mathfrak{n}$. Let Δ_0, the corresponding set of simple roots. Let $F = \{\alpha\}$, be with $\alpha\in\Delta_0$, and let (P_F, A_F), be the corresponding parabolic pair of G. Then is factible the Langlands canonical decomposition

$$P_F = {}^0M_FA_FN_F, \qquad \text{(II. 1.8)}$$

Let also ρ_F, defined by $\rho_F(H) = (1/2)\mathrm{tr}(\mathrm{ad}\ H|_{\mathfrak{a}P})\ \forall\ H\in\mathfrak{a}$. Let´s express to $f^P(am)$, $\forall$ $f\in C_c(K\backslash G/K)$, as

$$f^P(am) = a^{-\rho_F}\int_{N_F} f(nam)dn_F,$$

$\forall\ a\in A_F$, $m\in{}^0M_F$. In this sense, is necessary to take in count that dn_F, has been elected of invariant measure on N_F. We consider also that $^*P_F = P \cap M_F$, Then $^*P_F$, is a minimal parabolic subgroup of M_F, with the respective Langlands decomposition to $^*P_F$, to know,

$$^*P_F = {}^0MA^*N_F,\quad ^*N_F = N \cap M_F,$$

Normalizing the invariant measure, d^*n_F, on $^*N_F$, such that

$$dn = dn_F d^*n_F, \tag{II. 1.9}$$

Let's notice that $f^P \in C_c^\infty(K_F \backslash M_F/K_F)$. Let $^*F_g$, be denoting the "F_f" to M_F. Then is had that

$$F_f = {}^*F_g, \quad g = f^P, \tag{II. 1.10}$$

Now we can consider $s_\alpha H = Ad(k)H$, with $k \in K_F$. Therefore to demonstrate the proposition II. 1.1, only is enough demonstrate (II. 1. 10), under the set of endomorphic actions $s_\alpha H = Ad(k)H$, taking the case when $\Delta_0 = \{\alpha\}$. Let $p_0 = p_\gamma$, be to the class γ, of the trivial representation. Let β, be an automorphism of $U(\mathfrak{a})$, defined by

$$\beta(H) = H + \rho(H)\mathbf{1}, \quad \forall \; H \in \mathfrak{a},$$

Let $\gamma_0 = \beta \circ P_0$, be. Then γ_0, is called the Harish-Chandra homomorphism.

Lemma II. 1.1. The following two sentences are equivalent:

i. $\gamma_0(U(\mathfrak{g})^K) \subset U(\mathfrak{a})^W$ $(W = W(\mathfrak{g}, \mathfrak{a}))$,

ii. $\Xi_\mu = \Xi_{s\mu}$, $\forall \; s \in W$, $\mu \in (\mathfrak{a}_{\mathbb{C}})^*$.

Proof. If $T \in Hom_K(V_\gamma, (H^{\sigma, \mu})_K)$,and if $u \in U(\mathfrak{g})^K$, then

$$(uT)^\wedge = T^\wedge P_{\gamma, \sigma, \mu}(u),$$

In particular, $(H^{\sigma, \mu})_K$, have infinitesimal character of the form $\chi_{\Omega(\sigma, \mu)}$. This implies in special that $u \bullet \Xi_\mu = \mu(p_0(u))\Xi_\mu$, $\forall \; u \in U(\mathfrak{g})^K$, $\mu \in (\mathfrak{a}_C)^*$. If $u \in U(\mathfrak{g})a$, then

$$u\Xi_\mu(1) = Ad(k)u \bullet \Xi_\mu(\mathbf{1}).$$

Therefore from the analyticity of the functional Ξ_μ, the lemma is followed. ■

Lemma II. 1.2. If $\Delta_0 = \{\alpha\}$, then $\gamma_0(U(\mathfrak{g})^K) = \{h \in U(\mathfrak{a}) \mid s_\alpha h = h\}$.

Proof. Consider $\mathfrak{g}_I = [\mathfrak{g}, \mathfrak{g}]$. Then dim $\mathfrak{a} \cap \mathfrak{g}_I = 1$. Let $X_1, \ldots, X_p$, be an orthogonal base of $\mathfrak{p}_I = \mathfrak{p} \cap \mathfrak{g}_I$, relative to B. Let $C_\mathfrak{p} = \Sigma(X_j)^2$. For other side K, act transitively on the unitary sphere of $\mathfrak{p}_I$, and is clear that $S(\mathfrak{p})^K$, is the generated algebra for $\mathbf{1}$, $\mathfrak{p} \cap z(\mathfrak{g})$, and $C_\mathfrak{p}$. Then by the subquotient to $(\mathfrak{g}, K)$-modules, is known that

$$Ker \, \gamma_0 = U(\mathfrak{g})^K \cap U(\mathfrak{g})I_\gamma,$$

where

$$\gamma_0(U(\mathfrak{g})^K) = \gamma_0(Symm(S(\mathfrak{p}_{\mathbb{C}}))^K), \tag{II. 1.11}$$

A simple calculing show that exist constants c_1, and $c_2 \neq 0$, such that

$$\gamma_0(C_\mathfrak{p}) = c_1(H^2 + c_2),$$

The lemma is followed immediately. ■

Theorem II. 1.1. The following sucesion of algebra homomorphisms is exact:

$$0 \longrightarrow U(\mathfrak{g})^K \cap U(\mathfrak{g})I_\gamma \longrightarrow U(\mathfrak{g})^K \xrightarrow{\gamma_0} U(\mathfrak{a})^W \longrightarrow 0, \quad \text{(II. 1.12)}$$

With major pointity $\gamma_0 \circ \text{Symm} : S(\mathfrak{p}_C)^K \to U(\mathfrak{a})^W$, is a linear bijection.

We conclude this chapter with an estimation on the functional Ξ_μ, the which is used in the Casselman theorem.

Consider

$$\mathfrak{a}^+ = \{H \in \mathfrak{a} \mid \alpha(H) > 0, \forall \alpha \in \Phi^+\}, \quad \text{(II. 1.13)}$$

Let $A^+ = \exp \mathfrak{a}^+$, be. We consider furthermore that

$$G = K\mathrm{Cl}(A^+)K, \quad \text{(II. 1.14)}$$

which is true by the general theory of real reductive groups. Thus if $f \in C(K\backslash G/K)$, then f is completely determined for their values on $\mathrm{Cl}(A^+)$. Let

$$(\mathfrak{a}^*)^+ = \{\mu \in \mathfrak{a}^* \mid (\mu, \alpha) > 0, \forall \alpha \in \Phi^+\}, \quad \text{(II. 1.15)}$$

Let $W = W(\mathfrak{g}, \mathfrak{a})$, be. If $\mu \in \mathfrak{a}^*$, then there is an unique element in $W\mu \cap \mathrm{Cl}((\mathfrak{a}^*)^+)$. We use the notation $|\mu|$, to this element defined univocally. The theory of the real reductive groups implies that the intersection

$$W\mu \cap \mathrm{Cl}((\mathfrak{a}^*)^+) \neq \varnothing,$$

Therefore is possible to demonstrate that if $\beta, \sigma \in \mathrm{Cl}((\mathfrak{a}^*)^+)$, and if $s\beta = \sigma$, $\forall\, s \in W$, then $\sigma = \beta$. From lemma of Harish-Chandra explained in the general theory of (g, K)-modules, is deduced that $s\beta = \beta - Q$, with $Q = \Sigma c_\alpha$, the sum on $\alpha \in \Phi^+$, and whose coefficients $c_\alpha \geq 0$. Thus

$$(\beta, \beta) = (\sigma, \sigma) = (\beta - Q, \sigma) = (\beta, \sigma) - (\sigma, Q) \leq (\beta, \sigma) - (Q, \beta) \leq (\beta, \beta),$$

In particular $(Q, \beta) = 0$. But then $(\beta, \beta) = (\beta, \beta) + (Q, Q)$. Thus $Q = 0$.

Lemma II. 1.3. Let $\mu \in (\mathfrak{a}_C)^*$, then $|\Xi_\mu(a)| \leq a^{|\mathrm{Re}\mu|}\Xi_0(a)$, $\forall\; a \in \mathrm{Cl}(A^+)$.

Proof. From the integral functional expression $\Xi_\mu(a) = \int_K a(ka)^{\mu+\rho} dk$, is deduced that $|\Xi_\mu(g)| \leq \Xi_{\mathrm{Re}\mu}(g)$. Thus we can assume that $\mu \in \mathfrak{a}^*$. The proposition II. 1. 1, implies that we can assume that $\mu = |\mu|$. Let $a \in A$, be. Then

$$\Xi_\mu(a) = \int_K a(ka)^{\mu+\rho} dk = \int_N a(n)^{2\rho} a(k(n)a)^{\mu+\rho} dn, \quad \text{(II. 1.16)}$$

for the lemma to the invariant definition of dn [5], the measure on K. Then $k(n)\in Na(n)^{-1}n$. Thus

$$\Xi_\mu(a) = \int_N a(n)^{-\mu+\rho}\, a(na)^{\mu+\rho} dn, \tag{II. 1.17}$$

We note also that $a(na) = aa(a^{-1}na)$. Therefore has been demonstrated that

$$\Xi_\mu(a) = a^{\mu+\rho}\int_N a(n)^{-\mu+\rho}\, a(a^{-1}na)^{\mu+\rho} dn, \tag{II. 1.18}$$

If $a\in Cl(A^+)$, then considering the invariance of the measure dn, under the actions of the subgroup A, is obtained

$$(a(n)^{-1}a(a^{-1}na))^{\mu} \leq 1, \tag{II. 1.19}$$

Thus

$$\Xi_\mu(a) \leq a^{\mu}(a^{\rho}\int_N a(n)^{\rho}\, a(a^{-1}na)^{\rho} dn) = a^{\mu}\Xi_0(a), \tag{II. 1.20}$$

For (II. 1. 18). ■

Corollary II. 1.1. If $\mu\in(\mathfrak{a}_C)^*$, and from $a\in Cl(A^+)$, then $\Xi_\mu(a) \leq a^{|\mu|}$.

Proof. π_0, is an unitary representation of G (see elemental theory of representations in Hilbert space $L^2(G)$). Thus $\Xi_0 \leq 1$. The result is followed by the lemma II. 1. 3. ■

II. 2. Geometrical transforms: Integral transforms on Lie groups (Radon transforms on generalized flag manifolds)

Let G, be a connected reductive algebraic group over $\mathbb{C}$, and $\mathfrak{g}$, the Lie algebra of G. The group G, acts on $\mathfrak{g}$, by the adjoint action Ad. Let $\mathfrak{h}$, be a Cartan subalgebra of $\mathfrak{g}$, Δ, the root system in $\mathfrak{h}^*$, $\{\alpha_i | i\in I_0\}$, a set of simple roots Δ^+, the set of positive roots Δ^-, the set of negative roots $\mathfrak{h}^*_{\mathbb{Z}} = Hom(H, \mathbb{C}^\times) \subset \mathfrak{h}^*$, the weight lattice, and W, the Weyl group. For $\alpha\in\Delta$, we denote by $\mathfrak{g}_\alpha$, the corresponding root space and by $\alpha^\vee\in\mathfrak{h}$, the corresponding co-root. For $i\in I_0$, we denote by $s_i\in W$, the reflection corresponding to i. For $w\in W$, we set $l(w) = \#(w\Delta^-\cap\Delta^+)$. Set $\rho = \frac{1}{2}\sum_{\alpha\in\Delta^+}\alpha$, and define a (shifted) affine action of W, on $\mathfrak{h}^*$, by

$$w \circ \lambda = w(\lambda + \rho) - \rho, \tag{II. 2.1}$$

For $I \subset I_0$, we set

$$\Delta_I = \Delta \cap \sum_{i\in I}\mathbb{Z}\alpha_i, \quad \Delta_I^+ = \Delta_I\cap\Delta^+, \quad W_I = \langle s_i | i\in I\rangle \subset W,$$

$$\mathfrak{l}_I = \mathfrak{h} \oplus (\oplus_{\alpha\in\Delta_I}\mathfrak{g}_\alpha), \quad \mathfrak{n}_I = \oplus_{\alpha\in\Delta^+/\Delta_I}\mathfrak{g}_\alpha, \quad \mathfrak{p}_I = \mathfrak{l}_I\oplus\mathfrak{n}_I,$$

where

$$(\mathfrak{h}^*_{\mathbb{Z}})_I = \{\lambda \in \mathfrak{h}^*_{\mathbb{Z}} \mid \lambda(\alpha_i^\vee) \geq 0, \forall\, i \in I\},$$

$$(\mathfrak{h}^*_{\mathbb{Z}})^0_I = \{\lambda \in \mathfrak{h}^*_{\mathbb{Z}} \mid \lambda(\alpha_i^\vee) \geq 0, \forall\, i \in I\} \subset (\mathfrak{h}^*_{\mathbb{Z}})_I,$$

and also

$$\rho_I = (\Sigma_{\alpha \in \Delta + /\Delta I}\, \alpha) / 2,$$

We denote by w_I, the longest element of W_I. It is an element of W_I, characterized by $w_I(\Delta_I^-) = \Delta_I^+$. Let L_I, N_I, and P_I, be the subgroups of G, corresponding to $\mathfrak{l}_I$, $\mathfrak{n}_I$, and $\mathfrak{p}_I$.

For $\lambda \in (\mathfrak{h}^*_{\mathbb{Z}})_I$, let $V_I(\lambda)$, be the irreducible L_I-module with highest weight λ. We regard $V_I(\lambda)$, as a P_I-module with the trivial action of N_I, and define the generalized Verma module with highest weight λ, by

$$M_I(\lambda) = \mathcal{U}(\mathfrak{g}) \otimes_{\mathcal{U}(\mathfrak{p}_I)} V_I(\lambda), \tag{II. 2.2}$$

Let $L(\lambda)$, be the unique irreducible quotient of $M_I(\lambda)$ (note that $L(\lambda)$, does not depend on the choice of I, such that $\lambda \in (\mathfrak{h}^*_{\mathbb{Z}})_I$). Then any irreducible P_I-module is isomorphic to $V_I(\lambda)$, for some $\lambda \in (\mathfrak{h}^*_{\mathbb{Z}})_I$, and we have $\dim V_I(\lambda) = 1$, if and only if $\lambda \in (\mathfrak{h}^*_{\mathbb{Z}})^0_I$. Moreover, any irreducible $(\mathfrak{g}, P_I)$- module is isomorphic to $L(\lambda)$, for some $\lambda \in (\mathfrak{h}^*_{\mathbb{Z}})_I$.

Let

$$X_I = G/P_I,$$

be the generalized flag manifold associated to I.

Consider the category equivalence given in the following proposition to quasi-G-equivariant D_Z-modules:

Proposition II. 2.1. Assume that Z = G/H, where H, is a closed subgroup of G, and set $x = eH \in Z$.

i. The category of G-equivariant $\mathcal{O}_Z$-modules is equivalent to the category of H-modules via the correspondence $\mathcal{M} \rightarrow \mathcal{M}(x)$.
ii. The category of quasi-G-equivariant D_Z-modules is equivalent to the category of $(\mathfrak{g}, H)$-modules via the correspondence $\mathcal{M} \rightarrow \mathcal{M}(x)$.

The statement (i) is well-known (see [40]), and (ii) is due to Kashiwara [41].

Then the isomorphism classes of G-equivariant $\mathcal{O}_{X_I}$-modules (resp. quasi-G-equivariant D_{X_I}-modules) are in one-to-one correspondence with isomorphism classes of P_I-modules (resp.

($\mathfrak{g}$, P_I)-modules). For $\lambda \in (\mathfrak{h}^*_Z)_I$, we denote $\mathcal{O}_{X_I}(\lambda)$, the G-quivariant $\mathcal{O}_{X_I}$-module corresponding to the irreducible P_I-module $V_I(\lambda)$. We see easily the following.

Lemma II. 2.1. Let $\lambda \in (\mathfrak{h}^*_Z)_I$. The quasi-G-equivariant D_{X_I}-module corresponding to the ($\mathfrak{g}$, P_I)-module $M_I(\lambda)$, is isomorphic to $\mathcal{D}\mathcal{O}_{X_I}(\lambda) = \mathcal{D}_{X_I} \otimes_{\mathcal{O}_{X_I}} \mathcal{O}_{X_I}(\lambda)$.

We need the following relative version of the Borel-Weil-Bott theorem later (see Bott [42]).

Proposition II. 2.2. Let $I \subset J \subset I_0$, and let $\pi: X_I \to X_J$, be the canonical projection. For $\lambda \in (\mathfrak{h}^*_Z)_I$, we have the following.

i. If there exists some $\alpha \in \Delta_J$, satisfying $(\lambda + \rho - 2\rho_I)(\alpha^\vee) = 0$, then we have $R\pi_*(\mathcal{O}_{X_I}(\lambda)) = 0$.
ii. Assume that $(\lambda + \rho - 2\rho_I)(\alpha^\vee) \neq 0$, for any $\alpha \in \Delta_J$. Take $w \in W_J$, satisfying $(w(\lambda + \rho - 2\rho_I))(\alpha^\vee) > 0$, for any $\alpha \in \Delta_J^+$. Then we have

$$R\pi_*(\mathcal{O}_{X_I}(\lambda)) = \mathcal{O}_{X_J}(w(\lambda + \rho - 2\rho_I) - (\rho - 2\rho_J))[-l(w_J w) - l(w_I)].$$

Let $I, J \subset I_0$, with $I \neq J$. The diagonal action of G, on $X_I \times X_J$, has a finite number of orbits, and the only closed one $G(eP_I, eP_J)$, is identified with $X_{I \cap J} = G/(P_I \cap P_J)$. We can consider the correspondence to *D*-modules

$$X \leftarrow S \rightarrow Y, \tag{II. 2.3}$$

with $X = X_I$, $Y = X_J$, and $S = X$, then (II. 2. 3) take the form

$$X_I \xleftarrow{f} X_{I \cap J} \xrightarrow{g} X_J, \tag{II. 2.4}$$

and the Radon transform $R(\mathcal{D}\mathcal{O}_{X_I}(\lambda))$, for $\lambda \in (\mathfrak{h}^*_Z)_I$. Since f, and g, are morphisms of G-manifolds then the functor

$$R : D^b(\mathcal{D}_X) \rightarrow D^b(\mathcal{D}_Y),$$

with the correspondence rule

$$R(\mathcal{M}) = g_* f^{-1}(\mathcal{M}),$$

called the Radon transform on algebraic *D*-modules, induces a functor

$$R : D^b_G(\mathcal{D}_{X_I}) \rightarrow D^b_G(\mathcal{D}_{X_J}), \tag{II. 2.5}$$

Note that we have

$$\Omega_g \cong \mathcal{O}_{X_{I \cap J}}(\gamma_{I,J}), \quad \text{for } \gamma_{I,J} = \Sigma_{\alpha \in \Delta_J^+ \setminus \Delta_J} \alpha. \tag{II. 2.6}$$

II. 3. Radon transform of quasi-equivariant D-modules

Let $\lambda \in (\mathfrak{h}_Z^*)_I$. We describe our method to analyse $R(\mathcal{D}\mathcal{O}_{XI}(\lambda)) = g_* \underline{f}^{-1}(\mathcal{D}\mathcal{O}_{XI}(\lambda))$. By

$$(\underline{f}^{-1}(\mathcal{D}\mathcal{O}_{XI}(\lambda)))(e(P_I \cap P_J)) \cong \mathcal{D}\mathcal{O}_{XI}(\lambda))(eP_I) \cong M_I(\lambda), \qquad \text{(II. 3.1)}$$

the quasi-equivariant $D_{XI\cap J}$-module $\underline{f}^{-1}(\mathcal{D}\mathcal{O}_{XI}(\lambda))$, corresponds to the ($\mathfrak{g}$, $P_I \cap P_J$)-module $M_I(\lambda) = \mathcal{U}(\mathfrak{g}) \otimes_{\mathcal{U}(\mathfrak{p}_I)} V_I(\lambda)$, under the category equivalence given in proposition II. 2. 1. Set

$$\Gamma = \{x \in W_I \mid x \text{ is the shortest element of } W_{I\cap J}x\}, \qquad \text{(II. 3.2)}$$

and

$$\Gamma_x = \{x \in \Gamma \mid l(x) = k\}, \qquad \text{(II. 3.3)}$$

It is well-known that an element $x \in W_I$, belongs to Γ, if and only if $x^{-1}\Delta^+_{I\cap J} \subset \Delta^+_I$. This condition is also equivalent to

$$(x(\lambda + \rho))(\alpha^\vee) > 0, \ \forall \alpha \in \Delta^+_{I\cap J}, \qquad \text{(II. 3.4)}$$

In particular, we have $x \circ \lambda \in (\mathfrak{h}_Z^*)_{I\cap J}$, for $x \in \Gamma$.

By Lepowsky [43] and Rocha-Caridi [44] we have the following resolution of the finite dimensional $\mathfrak{l}_I$-module $V_I(\lambda)$:

$$0 \to N_n \to N_{n-1} \to \ldots \to N_1 \to N_0 \to V_I(\lambda) \to 0, \qquad \text{(II. 3.5)}$$

with $n = \dim \mathfrak{l}_I / \mathfrak{l}_I \cap \mathfrak{p}_J$, and

$$N_k = \bigoplus_{x \in \Gamma_k} \mathcal{U}(\mathfrak{l}_I) \otimes_{\mathcal{U}(\mathfrak{l}_I \cap \mathfrak{p}_J)} V_{I\cap J}(x \circ \lambda), \qquad \text{(II.3.6)}$$

By the Poincaré-Birkhoff-Witt theorem we have the isomorphi

$$\mathcal{U}(\mathfrak{l}_I) \otimes_{\mathcal{U}(\mathfrak{l}_I \cap \mathfrak{p}_J)} V_{I\cap J}(x \circ \lambda) \cong \mathcal{U}(\mathfrak{p}_I) \otimes_{\mathcal{U}(\mathfrak{p}_{I\cap J})} V_{I\cap J}(x \circ \lambda), \qquad \text{(II. 3.7)}$$

Of $\mathcal{U}(\mathfrak{l}_I)$-modules, where $\mathfrak{n}_{I\cap J}$, acts trivially on $V_{I\cap J}(x \circ \lambda)$. For other side, the action of $\mathfrak{n}_I$, on $\mathcal{U}(\mathfrak{l}_I) \otimes_{\mathcal{U}(\mathfrak{l}_I \cap \mathfrak{p}_J)} V_{I\cap J}(x \circ \lambda)$, is trivial. Indeed, by $[\mathfrak{p}_I, \mathfrak{n}_I] \subset \mathfrak{n}_I$, we have

$$\mathfrak{n}_I \mathcal{U}(\mathfrak{l}_I) = \mathcal{U}(\mathfrak{l}_I)\mathfrak{n}_I,$$

and hence

$$\mathfrak{n}_I(\mathcal{U}(\mathfrak{p}_I) \otimes_{\mathcal{U}(\mathfrak{p}_{I\cap J})} V_{I\cap J}(x \circ \lambda)) \subset \mathcal{U}(\mathfrak{p}_I)\mathfrak{n}_I \otimes V_{I\cap J}(x \circ \lambda) \subset \mathcal{U}(\mathfrak{p}_I) \otimes \mathfrak{n}_I V_{I\cap J}(x \circ \lambda) = 0,$$

for $\mathfrak{n}_I \subset \mathfrak{n}_{I\cap J}$. Thus we obtain the following resolution of the finite dimensional $\mathfrak{p}_I$-module $V_I(\lambda)$ (with trivial action of $\mathfrak{n}_I$):

$$0 \rightarrow N'_n \rightarrow N'_{n-1} \rightarrow \ldots \rightarrow N'_1 \rightarrow N'_0 \rightarrow V_I(\lambda) \rightarrow 0, \tag{II. 3.8}$$

with

$$N'_k = \bigoplus_{x\in\Gamma_k} \mathcal{U}(\mathfrak{p}_I) \otimes_{\mathcal{U}(\mathfrak{p}_{I\cap J})} V_{I\cap J}(x \circ \lambda),$$

By tensoring $\mathcal{U}(\mathfrak{g})$, to (II. 3. 8) over $\mathcal{U}(\mathfrak{p}_I)$, we obtain the following resolution of the $(\mathfrak{g}, P_{I\cap J})$-module $M_I(\lambda)$:

$$0 \rightarrow \tilde{N}_n \rightarrow \tilde{N}_{n-1} \rightarrow \ldots \rightarrow \tilde{N}_1 \rightarrow \tilde{N}_0 \rightarrow M_I(\lambda) \rightarrow 0, \tag{II. 3.9}$$

with

$$N'_k = \bigoplus_{x\in\Gamma_k} M_{I\cap J}(x \circ \lambda),$$

Since the quasi-G-equivariant $D_{X_{I\cap J}}$- module corresponding to the $(\mathfrak{g}, P_{I\cap J})$-module $M_{I\cap J}(x \circ \lambda)$ is $\mathcal{D}\mathcal{O}_{X_{I\cap J}}(x \circ \lambda)$, we have obtained the following resolution of the quasi-G-equivariant $D_{X_{I\cap J}}$-module $\underline{f}^{-1}(\mathcal{D}\mathcal{O}_{X_I}(\lambda))$,

$$0 \rightarrow \mathcal{N}_n \rightarrow \mathcal{N}_{n-1} \rightarrow \ldots \rightarrow \mathcal{N}_1 \rightarrow \mathcal{N}_0 \rightarrow \underline{f}^{-1}(\mathcal{D}\mathcal{O}_{X_I}(\lambda)) \rightarrow 0, \tag{II. 3.10}$$

with

$$\mathcal{N}_k = \bigoplus_{x\in\Gamma_k} \mathcal{D}\mathcal{O}_{X_{I\cap J}}(x \circ \lambda), \tag{II.3.11}$$

The following results consist in to investigate on the $D_{X_{I\cap J}}$-module $\underline{g}_*(\mathcal{D}\mathcal{O}_{X_{I\cap J}}(x \circ \lambda))$, for $x\in\Gamma$. We first remark that

$$\underline{g}_*(\mathcal{D}\mathcal{O}_{X_{I\cap J}}(x \circ \lambda)) = \mathcal{D}_{X_J} \otimes_{\mathcal{O}_{X_J}} Rg_*(\mathcal{O}_{X_{I\cap J}}(x \circ \lambda + \gamma_{I,J})), \tag{II. 3.12}$$

Indeed, by (II. 2.6) we have

$$\begin{aligned}
\underline{g}_*(\mathcal{D}\mathcal{O}_{X_{I\cap J}}(x \circ \lambda)) &= Rg_*(\mathcal{D}_{X_J \leftarrow X_{I\cap J}} \otimes^L_{\mathcal{D}_{X_{I\cap J}}} \mathcal{D}_{X_{I\cap J}} \otimes^L_{\mathcal{O}_{X_{I\cap J}}} \mathcal{O}_{X_{I\cap J}}(x \circ \lambda)) \\
&= Rg_*(\mathcal{D}_{X_J \leftarrow X_{I\cap J}} \otimes^L_{\mathcal{O}_{X_{I\cap J}}} \mathcal{O}_{X_{I\cap J}}(x \circ \lambda)) \\
&= Rg_*(g^{-1}\mathcal{D}_{X_J} \otimes_{g^{-1}\mathcal{O}_{X_J}} \Omega_g \otimes_{\mathcal{O}_{X_{I\cap J}}} \mathcal{O}_{X_{I\cap J}}(x \circ \lambda)) \\
&= \mathcal{D}_{X_J} \otimes_{\mathcal{O}_{X_J}} Rg_*(\Omega_g \otimes_{\mathcal{O}_{X_{I\cap J}}} \mathcal{O}_{X_{I\cap J}}(x \circ \lambda)) \\
&= \mathcal{D}_{X_J} \otimes_{\mathcal{O}_{X_J}} Rg_*(\mathcal{O}_{X_{I\cap J}}(x \circ \lambda + \gamma_{I,J})),
\end{aligned}$$

Lemma II. 3.1. Let $\lambda\in(\mathfrak{h}^*_{\mathbb{Z}})_I$, and $x\in\Gamma$.

i. If $(x(\lambda+\rho))(\alpha^\vee)=0$, for some $\alpha\in\Delta_J$, then we have $Rg_*(\mathcal{O}_{X_{I\cap J}}(x\circ\lambda+\gamma_{I,J}))=0$.
ii. Assume that $(x(\lambda+\rho))(\alpha^\vee)\neq 0$, for any $\alpha\in\Delta_J$. Take $y\in W_J$, satisfying $(yx(\lambda+\rho))(\alpha^\vee)>0$, for any $\alpha\in\Delta^+_J$. Then we have

$$Rg_*(\mathcal{O}_{X_{I\cap J}}(x\circ\lambda+\gamma_{I,J}))=\mathcal{O}_{X_J}((yx)\circ\lambda)[-(l(w_J y)-l(w_{I\cap J}))], \tag{II.3.13}$$

Proof. [45]. ■

Proposition II. 3.1. Let $\lambda\in(\mathfrak{h}^*_{\mathbb{Z}})_I$. Then there exists a family $\{\mathcal{M}(k)^\bullet\}_{k\geq 0}$, of objects of $D_G^b(\mathcal{D}_{X_J})$, satisfying the following conditions.

i. $\mathcal{M}(0)^\bullet\cong R(\mathcal{D}\mathcal{O}_{X_I}(\lambda))$.
ii. $\mathcal{M}(k)^\bullet=0$, for $k>\dim \mathfrak{l}_I/\mathfrak{l}_I\cap\mathfrak{p}_J$.
iii. We have a distinguished triangle

$$C(k)^\bullet\to\mathcal{M}(k)^\bullet\to\mathcal{M}(k+1)^\bullet\xrightarrow{+1}, \tag{II.3.14}$$

where

$$C(k)^\bullet=\bigoplus_{x\in\Gamma_k(\lambda)}\mathcal{D}\mathcal{O}_{X_J}((y_x x)\circ\lambda)[l(x)-m(x)], \tag{II.3.15}$$

Proof. For $0\leq k\leq\dim \mathfrak{l}_I/\mathfrak{l}_I\cap\mathfrak{p}_J$, define an object $\mathcal{N}(k)^\bullet$, of $D_G^b(\mathcal{D}_{X_{I\cap J}})$, by

$$\mathcal{N}(k)^\bullet=\left[\ldots\to 0\to\mathcal{N}_n\to\mathcal{N}_{n-1}\to\ldots\to\mathcal{N}_k\to 0\ldots\right], \qquad (II.3.16)$$ (II. 3.16)

Where $\mathcal{N}_j$, has degree $-j$ (see (II. 3. 10) and (II. 3. 11) for the notation). For $k>\dim \mathfrak{l}_I/\mathfrak{l}_I\cap\mathfrak{p}_J$, we set $\mathcal{N}(k)^\bullet=0$. By $\mathcal{N}(0)^\bullet\cong\underline{f}^{-1}(\mathcal{D}\mathcal{O}_{X_I}(\lambda))$, we have $g_*\mathcal{N}(0)^\bullet\cong R(\mathcal{D}\mathcal{O}_{X_I}(\lambda))$. Set

$\mathcal{M}(k)^\bullet=g_*\mathcal{N}(k)^\bullet$. Then the statements (i) and (ii) are obvius. Let us show (iii). Applying g_*, to the distinguished triangle

$$\mathcal{N}_k[k]\to\mathcal{N}(k)^\bullet\to\mathcal{N}(k+1)^\bullet\xrightarrow{+1},$$

we obtain a distinguished triangle

$$\underline{g}_*\mathcal{N}_k[k]\to\mathcal{M}(k)^\bullet\to\mathcal{M}(k+1)^\bullet\xrightarrow{+1},$$

For (II. 3.11) and (II. 3.12) and Lemma 2. 3.1, we have

$$\underline{g}_*\mathcal{N}_k=\bigoplus_{x\in\Gamma_k(\lambda)}\mathcal{D}\mathcal{O}_{X_I}((y_x x)\circ\lambda)[-m(x)],$$

The statement (iii) is proved. ∎

Theorem II. 3.1. Let $\lambda \in (\mathfrak{h}^*_{\mathbb{Z}})_I$.

i. We have

$$\sum_p (-1)^p [H^p(R(\mathcal{DO}_{X_J}(\lambda)))] = \sum_{x \in \Gamma(\lambda)} (-1)^{l(x)-m(x)} [\mathcal{DO}_{X_J}(y_x x) \circ \lambda], \tag{II. 3.17}$$

in the Grothendieck group of the category of quasi-G-equivariant $\mathcal{D}_{XJ}$-modules [23].

ii. If $\Gamma(\lambda) = \varnothing$, then $R(\mathcal{DO}_{XI}(\lambda)) = 0$.
iii. If $\Gamma(\lambda)$, consists of a single element x, then

$$R(\mathcal{DO}_{XI}(\lambda)) = \mathcal{DO}_{XJ}((y_x x) \circ \lambda)[l(x) - m(x)], \tag{II. 3.18}$$

iv. If $l(x) \geq m(x)$, for any $x \in \Gamma(\lambda)$, then we have $H^p(R(\mathcal{DO}_{XI}(\lambda))) = 0$, unless $p = 0$.
v. If $(\lambda + \rho)(\alpha^\vee) < 0$, for any $\alpha \in \Delta_J^+ \setminus \Delta_I$, then there exists a canonical morphism

$$\Phi: \mathcal{DO}_{XJ}((w_J w_{I \cap J}) \circ \lambda) \rightarrow H^0(R(\mathcal{DO}_{XI}(\lambda))), \tag{II. 3.19}$$

Moreover, Φ, is an epimorphism if $l(x) > m(x)$, for any $x \in \Gamma(\lambda) \setminus \{e\}$.

Proof. The statements (i), (ii) and (iii) are obvious from proposition 2. 3. 1. The statement (iv) follows from proposition 2. 3. 1, and the fact that any locally free $\mathcal{O}_X$-module $\mathcal{L}$, we have $H^p(R(D\mathcal{L})) = 0$ for any $p<0$. We assume that λ, satisfies the assumption in (v). Then we have $e \in \Gamma(\lambda)$, and $y_e = w_J w_{I \cap J}$. Hence (v) follows from proposition 2. 3. 1. ∎

Consider the following technical lemma, over classes of root spaces around of orbits with elements $y_e = w_J w_{I \cap J}$.

Lemma II. 3.2.

i. The map $W_J \times \Gamma \rightarrow W_J W_I((y, x) \mapsto yx)$, is bijective.
ii. For $\lambda \in (h^*_{\mathbb{Z}})_I$, we have

$$\{y_x x \mid x \in \Gamma(\lambda)\} = \{w \in W_J W_I \mid (w(\lambda + \rho))(\alpha^\vee) > 0, \text{ for any } \alpha \in \Delta^+_J\}, \tag{II. 3.20}$$

And we have

$$l(x) - m(x) = l(y_x) + l(x) - \#(\Delta^+_J \setminus \Delta_I) = l(y_x x) - \#(\Delta^+_J \setminus \Delta_I), \tag{II. 3.21}$$

Proof. [45]. ∎

For $\lambda \in (\mathfrak{h}^*_Z)_I$, we set

$$\Xi(\lambda) = \{w \in W_J W_I \mid (w(\lambda + \rho))(\alpha^{\vee}) > 0, \text{ for any } \alpha \in \Delta^{+}{}_J\}, \quad \text{(II. 3.22)}$$

Using the lemma 2. 3. 2, above we can reformulate the theorem 2. 3. 1, as follows:

Theorem II. 3.2. Let $\lambda \in (\mathfrak{h}^{*}{}_{Z})_I$.

i. We have

$$\sum_{p} (-1)^{p} [H^{p}(R(\mathcal{D}\mathcal{O}_{X_I}(\lambda)))] = (-1)^{\#(\Delta_J^{+}:\Delta_I)} \sum_{w \in \Xi(\lambda)} (-1)^{l(w)} [\mathcal{D}\mathcal{O}_{X_J}(w \circ \lambda)], \quad \text{(II. 3.23)}$$

in the Grothendieck group of the category of quasi-G-equivariant D_{XJ}-modules [23].

ii. If $\Xi(\lambda) = \varnothing$, then $R(\mathcal{D}\mathcal{O}_{XI}(\lambda)) = 0$.
iii. If $\Xi(\lambda)$, consists of a single element w, then
iv. $R(\mathcal{D}\mathcal{O}_{XI}(\lambda)) = \mathcal{D}\mathcal{O}_{XJ}(w \text{ o } \lambda)[l(w) - \#(\Delta^{+}{}_J \setminus \Delta_I)]$, (II. 3. 24)
v. If $l(w) \geq \#(\Delta^{+}{}_J \setminus \Delta_I)$, for any $w \in \Xi(\lambda)$, then we have $H^{p}(R(\mathcal{D}\mathcal{O}_{XI}(\lambda))) = 0$.
vi. If $(\lambda + \rho)(\alpha^{\vee}) < 0$, for any $\alpha \in \Delta^{+}{}_J \setminus \Delta_I$, then there exist a canonical morphism

$$\Phi : \mathcal{D}\mathcal{O}_{XJ}((w_J w_{I \cap J}) \text{o } \lambda) \rightarrow H^{0}(R(\mathcal{D}\mathcal{O}_{XI}(\lambda))), \quad \text{(II.3.25)}$$

Moreover Φ, is a epimorphism if $l(x) > \#(\Delta^{+}{}_J \setminus \Delta_I)$, for any $w \in \Xi(\lambda) \setminus \{w_J w_{I \cap J}\}$.

Chapter III

Orbital Integrals on Reductive Lie Groups

III. 1. Orbital integrals on reductive Lie groups [39]

Let G, be of inner type. Let

$$K'' = \{k \in K \mid \det(I - Ad(k)|_{\mathfrak{p}}) \neq 0\}, \tag{III. 1}$$

If $\mathfrak{h}$, is of Cartan belonging to $\mathfrak{g}$, be $H = \exp(\mathfrak{h})$. Let

$$H' = \{h \in H \mid \det(I - Ad(h)|_{\mathfrak{h}\perp}) \neq 0\}, \tag{III. 2}$$

Let

$$G[H'] = \{ghg^{-1} \in H \mid h \in H', g \in G\}, \tag{III. 3}$$

Then G[H'], is an open subset of G. Indeed, that be an open subset of G, it means that $G[H'] \subset Int(G)^{\perp}$. Indeed, $Int(G) \subset Aut(G) \subset G$. The relation of their closed subgroups is transitive in G. Then for definition

$$Int(G) = \{g \in G \mid dI\hat{g} = Ad(g)\}, \tag{III. 4}$$

But this is equal to that $\forall \hat{g} \in Int(G)$,

$$\det(I - Ad(\hat{g})|_{\mathfrak{h}\perp}) = 0, \tag{III. 5}$$

Then

$$Int(G)^{\perp} = \{g \in G \mid \det(I - Ad(\hat{g})|_{\mathfrak{h}\perp}) = 0\}, \tag{III. 6}$$

Since $ghg^{-1} = Ad(h)|_{\mathfrak{h}\perp} \forall\ g \in G$, then $G[H'] \subset Int(G)^{\perp}$.

Now, of the proof of lemma of the Weyl's formula is deduced that

$$\int_{G[H']} f(g)dg = (1/w) \int_H |\det((I - Ad(h)|_{\mathfrak{h}\perp})| \int_{G/H} f(ghg^{-1})\, dgH\, dh, \tag{III. 7}$$

where dg is an invariant measure on G, we fix an invariant measure on H, and we take to dgH, like the invariant quotient measure on G/H.

Let $w = o(N(H)/H)$, be where $N(H) = \{g \in G \mid ghg^{-1} = H\}$. Be $\mathfrak{h} = \mathfrak{t} \oplus \mathfrak{a}$, and $\mathfrak{g} = \mathfrak{t} \oplus \mathfrak{p}$. Let

$$\tau = \{H \in \mathfrak{h} \mid \det(ad(H)|_{\mathfrak{p}}) \neq 0\}, \tag{III. 8}$$

which is equal to **t**, a subalgebra of Cartan of $\mathfrak{g}$, that consider to T, also as

$$\mathrm{T} = \{t \in T \mid \det(\mathrm{Ad}(t)|_{\mathfrak{t}\perp}) \neq 0\}, \quad \text{(III. 9)}$$

of where K'' $\neq \varnothing$.

The previous integral formula applied to K, and G, implies that,

Lemma III. 1 Exist a positive constant C, such that

$$\int_{G[K'']} f(g)dg = c\int_K |\det((I - \mathrm{Ad}(k)|_{\mathfrak{p}})| \int_G f(gkg^{-1})\, dg\, dk, \quad \text{(III. 10)}$$

Proof. Indeed, consider you

$$G[K''] = \{gkg^{-1} \in H \mid k \in K'', g \in G\}, \quad \text{(III. 11)}$$

with

$$K'' = \{k \in K \mid \det(I - \mathrm{ad}(k)|_{\mathfrak{p}}) \neq 0\}, \quad \text{(III. 12)}$$

Since K, is compact then $N(K) \cong G[K'']$. Then $gKg^{-1} \subset G\ \forall\ g \in G$. Therefore,

$$|\det((I - \mathrm{Ad}(h)|_{\mathfrak{h}\perp})| \leq c|\det((I - \mathrm{Ad}(k)|_{\mathfrak{p}})|, \quad \text{(III. 13)}$$

Then by the integral formula of $\int_{G[K'']} f(g)dg$, is had that:

$$\int_{G[K'']} f(g)dg = \int_K |\det((I - \mathrm{Ad}(H')|_{\mathfrak{h}\perp})| \int_{G/K} f(gkg^{-1})\, dgK\, dk, \quad \text{(III. 14)}$$

and since $gKg^{-1} \subset G$, if and only if $G/K \cong G$, then $gK = g$, of where is obtained the conclusion of lemma.

If $\varepsilon > 0$, then we have

$$G_{e,\varepsilon} = \{gtg^{-1} \mid \det(\mathrm{Ad}(t) - I)|_{\mathfrak{t}\perp})| > \varepsilon\}, \quad \text{(III. 15)}$$

and

$$K_\varepsilon = \{k \in K \mid \det(\mathrm{Ad}(k) - I)|_{\mathfrak{p}})| > \varepsilon\}, \quad \text{(III. 16)}$$

and $(G'')_{e,\varepsilon} = G[K_\varepsilon]$.

In fact, the defined subgroups explicitly and previously are the spherical forms of the orbits G[K''], and K''.

Fixing a norm $\|,\|$, on G, the which we assume given like the corresponding norm operator to a representation (π, F), of G, on the Hilbert space of finite dimension F. Also we assume that $\pi(g)^* = \pi(\theta g)$, and that $\det\pi(g) = 1$, $\forall\ g \in G$.

Lemma III. 2 Let $0 < \varepsilon < 1$. Then

$$\int_{(G'')_{e,\varepsilon}} (1 + \log\|g\|)^{-d}\Xi(g)dg \leq C_d\varepsilon^{-d/2}\int_{A+}\gamma(a)\,(1 + \log\|a\|)^{-d}\Xi(a)da\,, \tag{III. 17}$$

where γ is such that $\forall\ a\in A$,

$$\gamma(a) = \prod_{\alpha\in IR} \text{senh}\,(\alpha(H)). \tag{III. 18}$$

Proof. The lemma. III. 1, affirms that under the constants of normalization, if f, is integrable on G, then

1. $$\int_G f(g)dg = \int_{K\times A+\times K} f(k_1ak_2)\gamma(a)\,dk_1da\,dk_2, \tag{III. 19}$$

This implies that $\forall\varepsilon > 0$,

$$\int_{(G'')_{e,\varepsilon}} (1 + \log\|g\|)^{-d}\Xi(g)dg =$$

$$=\int_{K_e}|\det((Ad(k) - I)|_{\mathfrak{p}})|\int_{K\times A+} (1 + \log\|au^{-1}kua^{-1}\|)^{-d}\bullet\Xi(\,au^{-1}kua^{-1})du\,da\,dk, \tag{III. 20}$$

Since $u^{-1}ku = k$, $\forall\ u\in K_\varepsilon$, with $\varepsilon > 0$, then

$$\int_{K_e}|\det((Ad(k) - I)|_{\mathfrak{p}})|\int_{K\times A+} (1 + \log\|au^{-1}kua^{-1}\|)^{-d}\bullet\Xi(\,au^{-1}kua^{-1})du\,da\,dk =$$

$$= \int_{K_e}|\det((I - Ad(k))|_{\mathfrak{p}})|\int_{A+} (1 + \log\|aka^{-1}\|)^{-d}\bullet\Xi(\,aka^{-1})\,da\,dk, \tag{III. 21}$$

If $X\in\mathfrak{p}$, then $\pi(X)$, is self adjunct. Indeed, if $X\in\mathfrak{p}$, then

$$(I - Ad(k)|_{\mathfrak{p}})X = 0, \tag{III. 22}$$

of where $\pi(adX) = \pi(X)$. Thus $\pi(X)$, is self adjunct. Then if $a\in A$, then $a = \exp H$, with $H\in\mathfrak{a}$. Thus,

$$\|aka^{-1}\|= \|aka^{-1}k^{-1}\|=\|\exp H\,\exp(-Ad(k)H)\|, \tag{III. 23}$$

Since $e^{[(I - Ad(k))H]} = e^{\pi(H)}\,e^{-Ad(k)H}$. But by Apendix A, if $\pi(X)$ is self adjunct then $tr\pi(X) = 0$, and $\dim F > 1$, then $tr[(I - Ad(k))H] = 0$, and due to that $\dim F > 1$, then

$$\|e^{\pi(H)}\,e^{-Ad(k)H}\|\geq e^{\|[(I - Ad(k))H]\|/(\dim F - 1)}, \tag{III. 24}$$

reason for which we see that

$$\log\|aka^{-1}\|\geq\|[\pi(H - Ad(k))H]\|/(\dim F - 1), \tag{III. 25}$$

We define $\forall\ k\in K$, the minimal value $\mu(k)$ like

$$\mu(k)=\min\|(Ad(k)- I)|_{\mathfrak{p}}\|, \tag{III. 26}$$

Then we have demonstrated that

2. ∃ a positive constant C, such that

$$\log\|aka^{-1}\| \geq C\mu(k)\log\|a\|, \tag{III. 27}$$

Let $\mu_1, \ldots, \mu_{2q}$, be the eigenvalues of $(I - Ad(k))|_{\mathfrak{p}}$, with numerable multiplicity. If we assume that $k \in K''$, then $(I - Ad(k))|_{\mathfrak{p}} \neq 0$, and since (III. 26) then $|\mu_j| = \mu(k), \forall\, j = 1, 2$. Then is clear that $|\mu_j| \leq 2, \ \forall\, i$, since

$$|\mu_j| = \mu(k) \leq \mathrm{tr}\, \pi(X)\, \pi(X)^* + \mathrm{tr}\, \pi(X)^{-1}\pi(X)^{-1*} = 1 + 1 = 2, \tag{III. 28}$$

since $\det \pi(X) = 1, \ \forall\, x \in \mathfrak{p}$. Thus, if $k \in K_\varepsilon$, then

$$\varepsilon < |\mu_1, \ldots, \mu_{2q}| \leq \mu(k)^2 2^{2q-2}, \tag{III. 29}$$

If $C = 2^{-q+1}$, then we have demonstrated that

3. If $k \in K_\varepsilon$, then$\mu(k) \geq C\varepsilon^{1/2}$.

This combined with (2) and the realized calculus to the begining implies that

$$\int_{(G'')_{e,\varepsilon}}(1 + \log\|g\|)^{-d}\Xi(g)dg \ \leq C^d\varepsilon^{1/2}\int_K|\det((Ad(k) - I)|_{\mathfrak{p}})|\int_A\gamma(a)(1 + \log\|a\|)^{-d}\bullet$$

$$\bullet\Xi(aka^{-1})da\, dk \leq C^d C^1\varepsilon^{(-d/2)}\int_{A^+}\gamma(a)(1 + \log\|a\|)^{-d}\int_K\Xi(aka^{-1})da\, dk, \tag{III. 30}$$

If $f \in C(G)$, then we define the function $f^{\sim}$,as the map

$$f^{\sim}: G \rightarrow N(G) \tag{III. 31}$$

with rule of corresponding

$$g \mapsto f^{\sim}(g) = \int_K f(kgk^{-1})dk, \tag{III. 32}$$

If $f \in C_c^{\infty}(G)$, then define the map Q_f, like

$$Q_f: K \rightarrow N(K) \tag{III. 33}$$

with rule of corresponding

$$k \mapsto Q_f(k) = \int_G f(gkg^{-1})dg, \tag{III. 34}$$

with domain on the space of the $k \in K$, to the which the integral converge absolutely.

Lemma III. 3 If $f \in C_c^{\infty}(G)$, then the domain of Q_f, it contains K''. In consequence $Q_f \in C^{\infty}(K'')$.

Proof. Let $h(g) = |f^{\sim}(g)|$. Then

$$\int_G h(gtg^{-1})dg = \int_G|f^{\sim}(gtg^{-1})|dg, \tag{III. 35}$$

But $|f^{\sim}(gtg^{-1})| \leq (1 + \log\|a\|)^{-d}\Xi(gtg^{-1})\sigma_{1,1,r}(f)$, with $\sigma_{1,1,r}(f) = 1$, since $f \in C_c^{\infty}(G)$. But by lemma III. 2, and considering $\gamma(a) = (1 + \log\|a\|)^{-d}$, with $d > 0$, we have that

$$\int_G (1 + \log\|a\|)^{-d}\Xi(gtg^{-1})dg = \int_{A+\times K} \gamma(a)\, \Xi(autu^{-1}a^{-1})da\, du, \qquad \text{(III. 36)}$$

Then

$$\int_G |f^{\sim}(gtg^{-1})|dg = \int_{A+\times K} \gamma(a)h(autu^{-1}a^{-1})da\, du, \qquad \text{(III. 37)}$$

Thus

$$\int_G h(gtg^{-1})dg = \int_{A+\times K} \gamma(a)h(autu^{-1}a^{-1})da\, du, \qquad \text{(III. 38)}$$

Then the argument of the demonstration in the lemma III. 2 (2), demonstrates that

4. If $k \in K_{\varepsilon}$, and if $a \in A$, then $\log|ata^{-1}| \geq \varepsilon^{1/2} C \log\|a\|$, which implies that if $u \in C_c^{\infty}(G)$ and if supp $u \subset B_r(G)$, where $B_r(G)$, is the full ball with $r > 0$,

$$B_r(G) = \{g \in G \mid \log\|g\| \leq r\}, \qquad \text{(III. 39)}$$

Then exist $c > 0$, independient of u, such that $u(autu^{-1}a^{-1}) = 1$, only into of the region

$B_r(G)$. Then the integral

$$\int_{A+\times K} \gamma(a)h(autu^{-1}a^{-1})da\, du < \infty, \qquad \text{(III. 40)}$$

$\forall k \in K''$, that is to say; $h(G_e, \varepsilon) \subset B_r(G)$, thus $K \supset K''$. Then the integral converge $\forall\, k \in K''$.

Now we demonstrates that $Q_f \in C^{\infty}(K'')$. Indeed, let $X_1, \ldots, X_n$, be a base of $\mathfrak{g}$. If $Y \in \mathfrak{t}$, then

$$Ad(g)Y = \Sigma C_j(g)X_j, \qquad \text{(III. 41)}$$

with each C_j, a matrix coefficient of the representation (Ad, g), of where

$$d^k/dt^k|_{t=0} f(gk\, \exp tYg^{-1}) = (Ad(g)Y)^k f(gkg^{-1}) = (\Sigma C_j(g)X_j)^k f(gkg^{-1}), \qquad \text{(III. 42)}$$

Thus exist constants $D > 0$, and $u \geq 0$, such that $|C_j(g)| \leq$ $D\|g\|^u$, of where

$$|d^k/dt^k|_{t=0} f(gk\, \exp tYg^{-1})| \leq C\|g\|^{ku}\Sigma|u_j f(gkg^{-1})|, \qquad \text{(III. 43)}$$

where u_j, is a base of $U^k(\mathfrak{g}_C)$. Newly this implies that if $f \in C^{\infty}(G)$, $\exists\, r > 0$, such that

$$|f(gkg^{-1})| \leq r, \qquad \text{(III. 44)}$$

$\forall\, g \in G$, and $k \in K``$, since $\Sigma|u_j f(gkg^{-1})| \leq \Sigma|u_j||f(gkg^{-1})| = \Sigma|u_j|r$. Then $Q_f \in C^{\infty}(K'')$.

We fix to Φ^+, on $\Phi(\mathfrak{g}_C, t_C)$. We make that ρ, is T-integral (which is possible due to that always is feasible to give a covering group of G).

Let $W = W(\mathfrak{g}_C, t_C)$, be then

$$\Delta(t) = t^{\rho}\prod_{\alpha\in\Phi^+}(1 - t^{-\alpha}) = \Sigma_{s\in W}\det(s)t^{s\rho}, \tag{III. 45}$$

$\forall\ t\in T$. Let $T'' = T \cap K''$. If $f\in C_c^{\infty}(G)$, then

5. $F_f^T(t) = \Delta(t)\int_G f(gtg^{-1})dg,\ \ \forall\ \ t\in T''$.
Indeed, since $Ad(G) \subset G$, and $gtg^{-1}\in gTg^{-1}\forall\ g\in G$. Then $Ad(G)t \subset gTg^{-1}\in N(T)$. Thus $\forall\ g\in G$, $Ad(g)t = t$, $\forall\ t\in T$, since the space of actions on the torus T, is

$$Ad(G)t = \{g\in G \mid Ad(g)t \subset T\ \forall\ t\in T\}, \tag{III. 46}$$

Then

$$\int_G f(Ad(g)t)dg = \int_G f(g)dg, \tag{III. 47}$$

since the left member of (III. 46) absolutely converge $\forall\ t\in T''$, and $f\in C_c^{\infty}(G)$. Also that,

$$\int_G f(g)dg = \pi(t)\int_G f(gtg^{-1})dg, \tag{III. 48}$$

But for the lemma lemma III. 1, and since

$$T'' = \{t\in T \mid \det(A(t)|_{\mathfrak{p}}) \neq 0\}, \tag{III. 49}$$

we have that

$$\int_G f(g)dg = \int_{T''}|\det(Ad(t)|_{\mathfrak{p}})|\int_G f(gtg^{-1})dgdt, \tag{III. 50}$$

But in $W = W(\mathfrak{g}_C, t_C)$, is satisfied that

$$\int_{T''}|\det(Ad(t)|_{\mathfrak{p}})|dt = \Sigma_{s\in W}\det(s)t^{s\rho} = \Delta(t), \tag{III. 51}$$

of where

$$F_f^T(t) = \Delta(t)\int_G f(gtg^{-1})dg, \tag{III.52}$$

$\forall\ t\in T''$.

Likewise, the lemma III. 1, implies that $F_f^T(t)\in C^{\infty}(T'')$. Consider

$$T_\varepsilon = G_{e,\varepsilon} \cap T = \{t\in T \mid |\Delta(t)|^2 > \varepsilon\}, \tag{III. 53}$$

which is equal$|\det(Ad(t)|_{\mathfrak{p}})| = |\pi_n(t)|^2$, $\forall$ $t \in T$. But $|\Delta(t)|^2 > \varepsilon$, $\forall$ $t \in T$, then $\det(Ad(t)|_{\mathfrak{p}}) \neq 0$, $\forall$ $t \in T$. Then $T_\varepsilon \subset T''$.

Lemma. III. 4 Let d, be such that

$$\int_{A^+} (1 + \log\|a\|)^{-d} \Xi(a)^2 \gamma(a) da < \infty, \qquad \text{(III. 54)}$$

Then exist positive constants C, and u, such that if $f \in C_c^\infty(G)$, then

$$\int_{T_\varepsilon} |F_f^T(t)|\, dt \leq C e^{-u} \sigma_{1,1,d}, \qquad \text{(III. 55)}$$

Proof. Let$\sigma = \sigma_{1,1,d}$,be and let

$$\int_{G/Ge,_\varepsilon \cap T} |f(g)|\, dg \leq \int_{T_\varepsilon} |F_f^T(t)|\, dt \leq \int_{T_\varepsilon} |\Delta(t)|^2 \int_G |f(gtg^{-1})| dg\, dt, \qquad \text{(III. 56)}$$

Now $|\Delta(t)| \geq C\varepsilon^{1/2}$, $\forall$ $t \in T_\varepsilon$. Thus

$$\int_{T_\varepsilon} |F_f^T(t)|\, dt \leq C\varepsilon^{-1/2} \int_{T_\varepsilon} |\Delta(t)| \int_G |f(gtg^{-1})| dg\, dt$$

$$\leq C\varepsilon^{-1/2}\, \sigma(f) \int_{T_\varepsilon} |\Delta(t)|^2 \int_G (1 + \log\|gtg^{-1}\|)^{-d} \Xi(gtg^{-1})^2 \gamma(gtg^{-1}) dt\, dg$$

$$= C\varepsilon^{-1/2}\, \sigma(f) \int_{Ge,_\varepsilon} (1 + \log\|g\|)^{-d} \Xi(g) dg$$

$$\leq C_d \sigma(f) \varepsilon^{-(d+1)/2} \int_{A^+} (1 + \log\|a\|)^{-2d} \Xi(a)^2 da$$

$$\leq C_d \sigma(f) \varepsilon^{-(d+1)/2} \int_{A^+} \gamma(a) (1 + \log\|a\|)^{-d} \Xi(a)^2 da,$$

By the lemma III. 1, if we take u= (d + 1)/2, then is followed (III. 54), since

$$|F_f^T(t)| \leq \sigma_{1,1,d}(f) \Xi(a) (1 + \log\|a\|)^{-d}, \qquad \text{(III. 57)}$$

Now well,

$$\int_K \Xi\, (aka^{-1}) dk = \Xi(a) \Xi(a^{-1}) = \Xi(a)^2, \qquad \text{(III. 58)}$$

Then using the precedent inequalities we have that

$$\int_{(G''_{e,\varepsilon})})(1 + \log\|g\|)^{-d}\, \Xi\, (g) dg = \int_{A^+} \gamma(a) (1 + \log\|a\|)^{-d} \Xi(a)^2 da, \qquad \text{(III. 59)}$$

If $f \in C(G)$, then we define the function $f^{\sim}$, as the map

$$f^{\sim}: G \rightarrow N(G), \qquad \text{(III. 60)}$$

with rule of correspondence

$$g \mapsto f^{\sim}(g) = \int_K f(kgk^{-1}) dk, \qquad \text{(III. 61)}$$

If $f\in C_c^{\infty}(G)$, then we define the map Q_f, as

$$Q_f : K \rightarrow N(K), \qquad \text{(III. 62)}$$

with rule of correspondence

$$Q_f : k \mapsto Q_f(k) = \int_G f(gkg^{-1})dg, \qquad \text{(III. 63)}$$

With domain on the space of the $k\in K$, to the which, the integral converges absolutely.■

Lemma. III. 5 If $f\in C_c^{\infty}(G)$, then the domain of Q_f, includes K''. In consequence $Q_f\in C^{\infty}(K'')$.

Proof. Let $h(g) = |f^{\sim}(g)|$. Then

$$\int_G h(gtg^{-1})dg = \int_G |f^{\sim}(gtg^{-1})|dg, \qquad \text{(III. 64)}$$

but $|f^{\sim}(gtg^{-1})| \leq (1 + \log\|a\|)^{-d}\Xi(gtg^{-1})\sigma_{1,1,r}(f)$, by lemma III. 2, with $\sigma_{1,1,r}(f) = 1$, since $f\in C_c^{\infty}(G)$.

If $f\in C(G)$, then we write (III. 60) and if $f\in C_c^{\infty}(G)$, then (III. 62). But by lemma. 3, and considering $\gamma(a) = (1 + \log\|a\|)^{-d}$, with $d > 0$, we have that

$$\int_G (1 + \log\|a\|)^{-d}\, \Xi\, (gtg^{-1})dg = \int_{A+\times K} \gamma(a)\Xi(autu^{-1}a^{-1})dadu, \qquad \text{(III. 65)}$$

Then

$$\int_G |f^{\sim}(gtg^{-1})|dg = \int_{A+\times K} \gamma(a)h(autu^{-1}a^{-1})dadu, \qquad \text{(III. 66)}$$

then

$$\int_G h(gtg^{-1})\, dg = \int_{A+\times K} \gamma(a)h(autu^{-1}a^{-1})dadu, \qquad \text{(III. 67)}$$

Therefore the argument of the demonstration in the lemma III. 2 (2), demonstrates that

6. If $k\in K_{\varepsilon}$, and if $a\in A$, then $\log|ata^{-1}|\geq\varepsilon^{1/2}C\log\|a\|$,which implies that if $u \in C_c^{\infty}(G)$, and if supp $u \subset B_{1,1,r}(G)$, where $B_{1,1,r}(G)$, (where $B_{1,1,r}(G) = B_r(G)$) is the full ball with $r > 0$.

Then exist $C > 0$, independent of u, such that $u(autu^{-1}a^{-1}) = 0$, to $k\in K''$, actually;

$h(G_{e,\varepsilon}) \subset B_r(G)$, thus $K \supset K''$. Thus the integral converge $\forall$ $k\in K''$.

Now we demonstrate that $Q_f\in C^{\infty}(K'')$. In effect, be $X_1, X_2, \ldots, X_n$, a base of $\mathfrak{g}$. If $Y\in\mathfrak{t}$, then $Ad(g)Y = \Sigma C_j(g)X_j$, with each C_j,a matrix coefficient of the representation (Ad, g), of where

$$d^k/dt^k|_{t=0} f(gk\, \exp tYg^{-1}) = (Ad(g)Y)^k f(gkg^{-1}) = (\Sigma C_j(g)X_j)^k f(gkg^{-1}),$$

Exist constants $D > 0$, and $u \geq 0$, such that $|C_j(g)| \leq D\|g\|^u$.

Note: $Ad(g) = dIg$, $\forall\ g \in G$ and $Ad(g)Y = d/dt|_{t=0} g_t K$,

thus

$$|d^k/dt^k|_{t=0}\ f(gk\ esptYg^{-1})| \leq C||g||^{ku} \Sigma |u_j f(gkg^{-1})|, \qquad \text{(III. 68)}$$

where u_j, is a base of $U^k(\mathfrak{g}c)$. Newly, this implies that if $f \in C_c^\infty(G)$, $\exists\ r > 0$, such that

$$|f(gkg^{-1})| \leq r,$$

$\forall g \in G$, and $\forall k \in K''$, since

$$\Sigma |u_j f(gkg^{-1})| \leq \Sigma |u_j||f(gkg^{-1})| \leq \Sigma |u_j|\ r,$$

Thus $Q_f \in C^\infty(K'')$. ■

We fix to Φ^+, on$\Phi(\mathfrak{g}c,\ tc)$. We assume that ρ, is T-integral (which is possible due to that always is easy to give a covering group of G).

Let $W = W(\mathfrak{g}c,\ tc)$, be then

$$\Delta(t) = t^\rho \prod_{\alpha \in \Phi^+} (1 - t^{-\alpha}) = \Sigma_{s \in W} \det(s) t^{s\rho}, \qquad \text{(III. 69)}$$

$\forall t \in T$. Let $T'' = T \cap K''$. If $f \in C_c^\infty(G)$, then

1. $$F_f^T(t) = \Delta(t) \int_G f(gtg^{-1}) dg, \quad \forall\ t \in T''.$$

Indeed, since $Ad(G) \subset G$ and $gtg^{-1} \in gTg^{-1}$, $\forall\ g \in G$. Then

$$Ad(G)t\ (\subset gTg^{-1}) \in N(T).$$

■

We can demonstrate that:

$$(1')\quad F_{zf}^T(t) = \gamma(z) F_f^T(t),$$

$\forall\ f \in C_c^\infty(G)$, $t \in T'$, and $z \in Z(\mathfrak{g})$. In effect, for definition

$$F_f^T(t) = \Delta(t) \int_G f(gtg^{-1}) dg,$$

$\forall\ f \in C_c^\infty(G)$, $t \in T''$. If $z \in Z(\mathfrak{g}c)$, then $\gamma(z)$, is a homomorphism of Harish-Chandra defined as

$$\gamma : Z(\mathfrak{g}c) \rightarrow U(\mathfrak{h}c), \qquad \text{(III. 70)}$$

with rule of correspondence

$$z \mapsto \Delta(\delta(z))\Delta^{-1} = P_h, \qquad \text{(III. 71)}$$

$\forall$ h$\in$H, such that $\delta(z) = \Delta^{-1}(\delta(z))\Delta$. Then $\Delta(t)\delta(z) = \gamma(z)\Delta(t)$. Thus

$$F_{zf}^{T}(t) = \Delta(t)\int_G zf(gtg^{-1})dg = \Delta(t)\int_G \delta(z)f(gtg^{-1})dg, \qquad \text{(III. 72)}$$

since to σ_μ, a character of a representation of finite dimension of G, is

$$z\sigma_\mu|_H = \delta(z)(\sigma_\mu|_H),$$

which is the same that

$$zf(gtg^{-1}) = \delta(z)f(gtg^{-1}),$$

Then (III. 71), take the form

$$\delta(z)\Delta(t)\int_G f(gtg^{-1})dg = \gamma(z)F_f^{T}(t).$$

$\forall$ $f\in C_c^{\infty}(G)$, $t\in T'$, and $z\in Z(\mathfrak{g}_C)$. ∎

Let $\alpha_1,\alpha_2,\ldots,\alpha_n$, be elements in Φ^+. Be $T' = \{t\in T \mid t^{\alpha_i} \neq 1, I = 1, 2, \ldots, r\}$. If $\varepsilon > 0$, then

$$T' = \{t\in T \,|\, |1 - t^{\alpha_i}| > \varepsilon, \ \forall \ i = 1, 2, \ldots, r\}, \qquad \text{(III. 73)}$$

Let

$$B(T') = \{f\in C^{\infty}(T') \,|\, \sigma_P(f) < \infty, \ \forall \ P\in U(t_C)\}, \qquad \text{(III. 74)}$$

Then is necessary show the following basic Harish-Chandra theorem:

Theorem III. 1.

1. If $f\in C_c^{\infty}(G)$, then $F_f^T\in B(T')$.
2. The map $f \mapsto F_f^T$, of $C_c^{\infty}(G)$ in $B(T')$, extends to a continuous map of $C(G)$, in $B(T')$.

Proof. Let $V = C(G)$, and $W = C_c^{\infty}(G)$, $s(w) = F_w$, $A = Z(\mathfrak{g}_C)$, $\Upsilon = \gamma$, and $\sigma = C_\sigma$. Then $s(f)\in B(T')$, $\forall$ $f\in C_c^{\infty}(G)$, and s, extends to a continuous map of C(G), in B(T').

Theorem III. 2. Let μ, be a continuous semi-norm on B(T'). Then exist $d \geq 0$, and a continuous semi-norm σ, on C(G), such that

$$\int_K \mu(F_R^{T}{}_{(kg)f}(f))dk \leq (1 + \log\|g\|)^d \Xi(g)\sigma(f), \qquad \text{(III. 75)}$$

to $f\in C(G)$, and $g\in G$.

For the previous theorem III. 1, exist a continuous semi-norm σ, of C(G), and $q, d > 0$, such that if $f\in C(G)$, then

$$\int_K\int_{T_\varepsilon} |F_R^{T}{}_{(kg)f}(f))| dtdk \leq \varepsilon^{-q}\sigma(f)\,(1 + \log\|g\|)^d \Xi(g), \quad \forall \varepsilon > 0, \qquad \text{(III. 76)}$$

For the lemma. III. 5,

$$\int_K\int_{T_\varepsilon}|F_R{}^T{}_{(kg)f}(f))|dtdk = \int_K\int_{T_\varepsilon}|\Delta(T))|\int_G|f(xtx^{-1}kg))|\ dxdtdk,$$

of where $\forall\sigma$, on C(G), and $\varepsilon > 0$, since $|\Delta(t)|\geq C\varepsilon^{1/2}$, $\forall\ t\in T_\varepsilon$:

$$\int_K\int_{T_\varepsilon}|\Delta(T))|\int_G|f(xtx^{-1}kg))|\ dxdtdk \leq\sigma(f)\varepsilon^{1/2}\int_K\int_{T_\varepsilon}|\Delta(T))|^2(1 + \log||xtx^{-1}kg||)^d\bullet\Xi(xtx^{-1}kg)dxdtdk,$$

with $\sigma = \sigma_{1,\,1,\,d}$.

Note that the norm in G, satisfies $\forall$ x, y$\in$G, that

$$||xy||\geq||x||||y^{-1}||^{-1} = ||y||^{-1}||x||,$$

Thus $\log||xy|| + \log||y||\geq \log||x||$.Of where

$$(1 + \log||xy||)(1 + \log||y||) \geq 1 + \log||x||,$$

Thus the expression (III. 75), take the form

$$\int_K\int_{T_\varepsilon}|F_R{}^T{}_{(kg)f}(f))|dtdk \leq\varepsilon^{-1/2}\sigma(f)\ (1 + \log||g||)^d\int_{K\times T_\varepsilon}|\Delta(t)|^2\int_G(1 + \log||xtx^{-1}kg||)^{-d}\bullet\Xi(xtx^{-1}kg)dxdtdk \leq$$

$$\leq\varepsilon^{-1/2}\sigma(f)\Xi(g)(1 + \log||g||)^d\int_{G_{e,\,\varepsilon}} (1 + \log||x||)^{-d}\Xi(x)dg,$$

Then $\varepsilon^{-1/2}\sigma(f)\ (1 + \log||g||)^d\int_{T_\varepsilon}|\Delta(t)|^2\int_G(1 + \log||xtx^{-1}||)^{-d}\Xi(xtx^{-1}kg)dxdtdk$, but

$$\Xi(xtx^{-1}kg) = \Xi(xtx^{-1})\ \Xi(kg) = \Xi(xtx^{-1})\ \Xi(g),$$

then

$$\varepsilon^{-1/2}\sigma(f)\ (1 + \log||g||)^d\int_{T_\varepsilon}|\Delta(t)|^2\int_G(1 + \log||xtx^{-1}||)^{-d}\Xi(xtx^{-1}kg)dxdtdk =$$

$$= \varepsilon^{-1/2}\sigma(f)\ (1 + \log||g||)^d\int_{T_\varepsilon}|\Delta(t)|^2\int_G(1 + \log||xtx^{-1}||)\Xi(xtx^{-1})\ \Xi(g)\int_K dx,$$

Considering $G\cap T_\varepsilon = G_{e,\,\varepsilon}$, ($\varepsilon > 0$), and given that

$$dxdtdk = \mu(F^T{}_{R(kg)f})dk,$$

$\forall$ f$\in$C(G), and g$\in$G, then

$$\varepsilon^{-1/2}\sigma(f)\ (1 + \log||g||)^d\int_{T_\varepsilon}|\Delta(t)|^2\int_G(1 + \log||xtx^{-1}||)\Xi(xtx^{-1})\ \Xi(g)\int_K dx \leq$$

$$\leq\varepsilon^{-1/2}\sigma(f)\Xi(g)(1 + \log||g||)^d\int_{G_{e,\,\varepsilon}} (1 + \log||xtx^{-1}||)^{-d}\Xi(x)dg,$$

Then

$$\int_K\mu(F^T{}_{R(kg)f})dk \leq (1 + \log||g||)^d\sigma(f)\Xi(g).\ \blacksquare$$

Suppose that every subgroup of Cartan, ρ, is integral. Let $\mathfrak{h}$, be a subalgebra of Cartan of $\mathfrak{g}$, and let $H = \exp(\mathfrak{h})$. Let $\bar{\alpha}(h) = \overline{\alpha(\bar{h})}$, $\forall\ \alpha\in\Phi(\mathfrak{g}_C, \mathfrak{h}_C)$, $h\in\mathfrak{h}$, (here $\bar{X}$,is the conjugate complex of $X\in\mathfrak{g}_C$, relative to $\mathfrak{g}$). Let Φ^+, be a positive root system to $\Phi(\mathfrak{g}_C, \mathfrak{h}_C)$, such that if $\alpha\in\Phi^+$, and $\bar{\alpha} \neq -\alpha$, then $\alpha\in\Phi^+$. Let

$$\Sigma = \{\alpha \in \Phi^+ \mid \bar{\alpha} \neq -\alpha\}, \qquad \text{(III. 77)}$$

Let

$$\Delta_H(h) = h^\rho \prod_{\alpha\in\Phi^+-\Sigma} (1-h^{-\alpha}) \left| \prod_{\alpha\in\Sigma} (1-h^{-\alpha}) \right|, \qquad \text{(III. 78)}$$

Clearly $\Delta_H(h) = \pm\,\Delta(h)$, $\forall\ h\in H$. Indeed, for definition

$$\Delta(h) = h^\rho \prod_{\alpha\in\Phi^+} (1-h^{-\alpha}),$$

$\forall h\in H$. Since $(\Phi^+ - \Sigma) \cup \Sigma = \Phi^+$, then

$$\Delta_H(h) = h^\rho \left| \prod_{\alpha\in\Phi^+} (1-h^{-\alpha}) \right|,$$

which is equivalent to that

$$\left| \prod_{\alpha\in\Phi^+} (1-h^{-\alpha}) \right| = \begin{cases} -\prod_{\alpha\in\Phi^+} (1-h^{-\alpha}), & \text{if} \quad \alpha\in\Sigma, \\ +\prod_{\alpha\in\Phi^+} (1-h^{-\alpha}), & \text{if} \quad \alpha\in\Phi^+ - \Sigma, \end{cases}$$

Thus

$$\Delta_H(h) = h^\rho(\pm \prod_{\alpha\in\Phi^+} (1-h^{-\alpha})) = \pm h^\rho \prod_{\alpha\in\Phi^+} (1-h^{-\rho}) = \pm\Delta(h),$$

Now well, if $f\in C(G)$, then

2.
$$F_f^H(h) = \Delta_H(h) \int_{G/H} f(ghg^{-1})dgH,$$

The measure on G/H, is elected like in the theorem III. 1, and the domain is the set of all the $h\in H$, such that the integral converges absolutely.

We assume that $H = H_F = T_F A_F$, with (P_F, A_F), a parabolic cuspidal canonical pair. Assume that, like in theorem III. 1, we have

3.
$$F_f^H(h) = \Delta_H(h) \int_{K\times^0 M\times N} f(kmnhn^{-1}m^{-1}k^{-1})dkdmdn.$$

We define to $h\in H$, $\Gamma_h(n) = h^{-1}nhn^{-1}$, $\forall$ $n\in N_F$.

Likewise, is had the following lemma:

Lemma III. 5.

1. $\Gamma_h(N_F) \subset N_F$.
2. If $\det((Ad(h^{-1}) - I)|_{n_F}) \neq 0$, then Γ_h, is a diffeomorphism of N_F, in N_F.

In consequence, if f, is integrable on N_F, then

$$|\det((Ad(h^{-1}) - I)|_{n_F})| \int_{N_F} f(h^{-1}nhn^{-1})dn = \int_{N_F} f(n)dn, \qquad \text{(III. 79)}$$

The demonstration of this result is essentially the same that the realized by corresponding to the lemma IV. 2, that in the next chapter will be seen.

If $f\in C(G)$, let $\underline{f}$, as (III. 32)-(III. 34). We note that if $h\in H$, then

$$|\det((Ad(h^{-1}) - I)|_{n_F})| = |\Pi_{\alpha\in\Sigma}(1 - h^{-\alpha})|, \qquad \text{(III. 80)}$$

Then we can define the cusp forms:

$$F_f^H(h) = C_F h^{\rho}\Delta_M(h)\int_{^0M\times N_F} f(mhm^{-1}n)dmdn = C_F h^{-\rho}\Delta_M(h)\int_{^0M\times N_F} f(nmhm^{-1})dmdn, \qquad \text{(III. 81)}$$

Where Δ_M, is the "Δ", for $\Phi^+\cap\Phi(\mathfrak{m}_C, \mathfrak{h}_C)$. In the next chapter (III. 81) will appears again as other version adequated to Lie algebras.

Chapter IV

Orbital Integrals on Reductive Lie Algebras

IV. 1. Orbital integrals on reductive Lie algebras

Let $\mathfrak{t}'' = \mathfrak{t} \cap \mathfrak{T}''$, be where

$$\mathfrak{T}'' = \{X \in \mathfrak{T} \mid \det(\mathrm{ad}(X)|_{\mathfrak{p}}) \neq 0\}, \quad \text{(IV. 1)}$$

We fix a system of positive roots Φ^+,in$\Phi(\mathfrak{g}_C, \mathfrak{t}_C)$. Let

$$\pi = \prod_{\alpha \in \Phi^+} \alpha, \quad \text{(IV. 2)}$$

Let

$$\Phi_n = \{\alpha \in \Phi(\mathfrak{g}_C, \mathfrak{t}_C) \mid (\mathfrak{g}_C)_\alpha \subset \mathfrak{p}_C\}, \quad \text{(IV. 3)}$$

and let

$$\pi_n = \prod_{\alpha \in \Phi^+ \cap \Phi_n} \alpha, \quad \text{(IV. 4)}$$

If $H \in \mathfrak{t}$, then $|\det(\mathrm{ad}(X)|_{\mathfrak{p}})| = |\pi_n(H)|^2$. Let T, be the subgroup of Cartan of G, corresponding to $\mathfrak{t}$. Let

$$\mathfrak{S}(\mathfrak{g}) = \{f \in C^\infty(G) \mid \mu_{s,t}(f) < 0\}, \quad \text{(IV. 5)}$$

where

$$\mu_{s,t}(f) = \sup_{X \in \mathfrak{g}} \|X\|^r \Sigma |\partial^{|I|}/\partial x^I f(X)| < \infty, \quad \text{(IV. 6)}$$

where the topology of the space $\mathfrak{S}(\mathfrak{g})$, is the constructed by the semi-norms $\mu_{s,t}$. Note that the semi-norms $q_{s,t}$, are continuous on the space $\mathfrak{S}(\mathfrak{g})$.

If $f \in \mathfrak{S}(\mathfrak{g})$, and $H \in \mathfrak{t}''$, then we consider

$$\Phi_f^T(H) = \pi(H) \int_G f(\mathrm{Ad}(g)H)\, dg, \quad \text{(IV. 7)}$$

then $\Phi_f^T \in C^\infty(\mathfrak{t}'')$.

To demonstrate this, we elect $H_1, \ldots, H_r$, a orthonormal base of $\mathfrak{t}$, and let $t_1, \ldots, t_r$, be the corresponding coordinates in the algebra $\mathfrak{t}$.

Then is necessary proving that:

Lemma IV. 1 Exist a constant u, such that if I, is a rth-mult-index then exist a continuous semi-norm μ_I, on $\mathfrak{S}(\mathfrak{g})$, such that

$$|\partial^{|I|}/\partial x^{|I|}\, \Phi_f{}^T(H)| \leq |\pi_n(H)|^{-u}\mu_I(f). \qquad \text{(IV. 8)}$$

$\forall$ $f\in\mathfrak{S}(\mathfrak{g})$, $H\in\mathfrak{t}''$.

Proof. If $p\in S(\mathfrak{g}c)$, let $p^{\sim}$, be the image of isomorphism

$$S(\mathfrak{g}c) \rightarrow S(\mathfrak{t}c), \qquad \text{(IV. 9)}$$

with rule of correspondence

$$p \mapsto p^{\sim}, \qquad \text{(IV. 10)}$$

let the space

$$I = \{p^{\sim} \mid p\in S(\mathfrak{g}c)^G\}, \qquad \text{(IV. 11)}$$

Then by the theorem of the isomorphism of Harish-Chandra exist $p^{\sim}\in S(\mathfrak{g}c)$, such that $p^{\sim} = p|_{\mathfrak{t}c}$, whose linear isomorphism $\forall$ $p_j\in S(\mathfrak{g}c)^G$, is defined by

$$p^{\sim}_j = \mathrm{Res}_{S(\mathfrak{g}c/\mathfrak{t}c)}(p_j), \qquad \text{(IV. 12)}$$

with j = 1, 2, …, d. Then S(tc), is invariant under the module class I, that is to say, is a finitely generated module as I-module. Then $\forall$ $p_j\in S(\mathfrak{g}c)$, then

$$S(\mathfrak{t}c) = \Sigma_j I p_j, \qquad \text{(IV. 13)}$$

with j = 1, 2, …, d. We consider $p_j\in S(\mathfrak{t}c)$, j = 1, 2, …, d, such that (IV. 13). Be

$$\mathfrak{t}' = \{H\in\mathfrak{t} \mid \pi(H) \neq 0\}. \qquad \text{(IV. 14)}$$

Be $H\in\mathfrak{t}'$ and be $W = U_{\mathfrak{t}'}(H)$, the open neighborhood of H, in $\mathfrak{t}'$. Be U = Ad(G)W. If $f\in\mathfrak{S}(\mathfrak{g})$, then

$$g(X) = \int_G f(\mathrm{Ad}(g)X)dg, \qquad \text{(IV. 15)}$$

Then $g\in C^\infty(U)$.

We consider $Y\in \mathrm{Ad}(G)W = U$, then we define

$$g(\mathrm{Ad}(x)Y) = \int_G f[\mathrm{Ad}(x)(\mathrm{Ad}(x)Y))dx = \int_G f(\mathrm{Ad}(x)Y)dx = g(Y), \qquad \text{(IV. 16)}$$

since $Y = Ad(x)Y$, $\forall\, x\in G$. Then $g(Ad(x)Y) = g(Y)$, $\forall\; Y\in U$. From theorem III. 1, we have that $p\in S(\mathfrak{g}_C)^G$,

$$pg|_W = \pi^{-1}p^{\sim}\pi g|_W, \tag{IV. 17}$$

This implies that

$$p^{\sim}\Phi_f{}^T(H) = \Phi_{pf}{}^T(H), \tag{IV. 18}$$

$\forall\, f\in\mathfrak{S}(\mathfrak{g})$, and $H\in t'$. In effect, $\Leftarrow$):

$$\Phi_{pf}{}^T(H) = \pi(H)\int_G pf(Ad(g)H)dg$$

$$= \pi(H)\int_G \pi^{-1}p^{\sim}\pi f(Ad(g)H)dg$$

$$= \pi(H)\int_G p^{\sim}f(Ad(g)H)dg$$

$$= p^{\sim}\{\pi(H)\int_G f(Ad(g)H)dg\} = p^{\sim}\Phi_f{}^T(H),$$

$\forall\, H\in t'$, and $f\in\mathfrak{S}(G)$. $\Rightarrow$):

$$p^{\sim}\Phi_f{}^T(H) = p^{\sim}\{\pi(H)\int_G f(Ad(g)H)dg\}$$

$$= \pi(H)\int_G \pi^{-1}p^{\sim}\pi f(Ad(g)H)dg$$

$$= \pi(H)\int_G p|_{t'}f(Ad(g)H)dg$$

$$= = \pi(H)\int_G pf(Ad(g)H)dg = \Phi_{pf}{}^T(H),$$

$\forall\, H\in t'$, and $f\in\mathfrak{S}(G)$. If we consider as partial derivative of multi-index order, the product of the canonical components of the field $H\in t'$, we have

$$H_1{}^{i_1}, \ldots, H_r{}^{i_r} = \partial^{|I|}/\partial t^I = \Sigma_j u^{\sim}{}_j p_j, \tag{IV. 19}$$

$\forall\, u_j\in S(\mathfrak{g}_C)^G$. Then $\forall\, p_j\in S(t_C)$,

$$|p_j\Phi_f{}^T(H)| \leq c|\pi_n(H)|^{-r}\mu_j(f), \tag{IV. 20}$$

with μ_j, a continuous semi-norm on $\mathfrak{S}(\mathfrak{g})$, and r_j, depends only on the degree of p_j. Realizing an adequate election of μ_j, we can to replace r_j, for r, the maximum of the r_j. Thus

$$|p_j\Phi_f{}^T(H)| \leq c|\pi_n(H)|^{-r}\Sigma\mu_j(f), \tag{IV. 21}$$

$\forall\, H\in t'$. Thus

$$|\partial^{|I|}/\partial t^{I}\,\Phi_f{}^{T}(H)| \leq c|\pi_n(H)|^{-r}\Sigma\mu_j(u_j f). \quad \text{(IV. 22)}$$

$\forall$ $H\in t'$. Since both sides of the previous inequality are continuous on t'', then is followed (IV. 8). ■

If U, is an open of **t**, then we define to the space

$$\mathfrak{S}(U) = \{f\in C^{\infty}(U)\,|\,\mu_{U,\,r,\,s}(f) = \sup_{X\in U}||X||^{r}\Sigma_{|I|\leq s}|\partial^{|I|}/\partial t^{I}f(X)| < \infty\}, \quad \text{(IV. 21)}$$

Said space is a Frèchet space.

Theorem IV. 1(Theorem of Harish-Chandra). If $f\in\mathfrak{S}(\mathfrak{g})$, then $\Phi_f{}^{T}\in\mathfrak{S}(t'')$. Furthermore, the map $f\,|\!\rightarrow\Phi_f{}^{T}$, from $\mathfrak{S}(\mathfrak{g})$, to $\mathfrak{S}(t'')$ is continuous.

Proof. Let C, the convex component of t'', that is to say, the space

$$C = \{H\in t''\,|\,\pi_n(H) = \det(ad(H)|_{\mathfrak{p}})\}, \quad \text{(IV. 22)}$$

If $H\in$ i**t**, then $\alpha(H)\in\mathbb{R}$, $\forall\alpha\in\Phi$, $\Phi = \Phi(\mathfrak{g}_C, t_C)$. Thus, if $\alpha\in\Phi_n$, then $i\alpha > 0$, or $i\alpha < 0$, on the space C. Then we can to define the rule of correspondence

$$|\alpha|_C = \begin{cases} i\alpha, \text{ if } \alpha\geq 0, \\ -i\alpha, \text{ if } \alpha\geq 0, \end{cases} \quad \text{(IV. 23)}$$

Thus

$$|\pi_n(H)| = |\Pi_{\alpha\in\Phi^{+}\cap\Phi^{n}}\alpha|_C \leq \Pi_{\alpha\in\Phi^{+}\cap\Phi^{n}}|\alpha|_C(H), \quad \text{(IV. 24)}$$

$\forall$ $H\in$ i**t**. Let $x\in Cl(C)$, and we fix $x_0\in C$. Then

$$|\alpha|_C(x + tx_0) = |\alpha|_C(x) + t|\alpha|_C(x_0) \geq t|\alpha|_C(x_0), \quad \text{(IV. 25)}$$

if $t\geq 0$. Let $f\in\mathfrak{S}(\mathfrak{g})$. The previous result implies that if $F = \Phi_f$, and if $q = |\Phi^{+}\cap\Phi_n|$, then

$$|pF(x + tx_0)| \leq t^{-u_q}\mu_p(f), \quad \text{(IV. 26)}$$

$\forall$ $t > 0$, and $p\in\mathfrak{S}(t)$. μ_p is a continuous semi-normon $\mathfrak{S}(\mathfrak{g})$. In effect,

$$|\partial^{|I|}/\partial t^{I}\,\Phi_f(F)| \leq |\pi_n(x)|^{-u}\mu_I(f),$$

But

$$|\partial^{|I|}/\partial t^{I}\,\Phi_f(F)| \leq |d/dt\; pF(x + tx_0)| \leq |\pi_n(x)|^{-u}\mu_p(f), \quad \text{(IV. 27)}$$

$\forall$ $x\in Cl(C)$. Then

$$|pF(x + tx_0)| \leq |\pi_n(x)|^{-u}\mu_I(f) \leq t^{-uq}\mu_P(f), \quad \text{(IV. 28)}$$

$\forall\ x \in Cl(C)$, or $|pF(x + tx_0)| \leq t^{-uq}\mu_P(f)$, $\forall\ q = |\Phi^+ \cap \Phi_n|$.

Now, to the kth-derivative we have that

$$d^k/dt^k\, pF(x + tx_0) = (x_0)^k\, pF(x + tx_0), \quad \text{(IV. 29)}$$

$\forall\ x \in Cl(C)$, where is necessary have always all the time that:

$$S(t) = \{p_j \in S(\mathfrak{g}c) \mid p = \Sigma_j I p_j\}, \quad \text{(IV. 30)}$$

where $I = \{p^{\sim} \mid p \in S(\mathfrak{g}c)^G\}$. S(t), is identified with the space of differential operators of constants coefficients on the algebra t. This means that if $u(t) = pF(x + tx_0)$, then

$$u^{(k)}(t) = (x_0)^k pF(x + tx_0) \quad \text{(IV. 31)}$$

of where the value of the kth-derivative of u(t), comes bounded as

$$|u^{(k)}(t)| \leq t^{-uq}\mu_{x_P}(f)|\pi_n(x_0)|^{-u}, \quad \text{(IV. 32)}$$

Schollium: Let $u \in C^\infty([0, 1])$, be and suppose that

$$|u^{(k)}(t)| \leq t^{-m}a_k, \quad \text{(IV. 33)}$$

to $0 < t \leq 1$, and rank $k = 0, 1, 2, \ldots$. Then

$$|u(t)| \leq C(a_0 + \ldots + a_{m+1}), \quad \text{(IV. 34)}$$

to $0 < t \leq 1$. Here C, depends only on m. Indeed, we assume that $m \geq 1$. If $m > 1$, then we write to $u^{(k)}(t)$, as:

$$u^{(k)}(t) = \int_0^1 u^{(k+1)}(s)ds - \int_t^1 u^{(k+1)}(s)ds = u^{(k)}(1) - \int_t^1 u^{(k+1)}(s)ds, \quad \text{(IV. 35)}$$

Having that

$$|u^{(k)}(t)| \leq -\int_t^1 t^{-m}a_{k+1}dt + \int_0^1 t^{-m}a\_{k+1}dt + a_k, \quad \text{(IV. 36)}$$

Then

$$|u^{(k)}(t)| \leq a_{k+1}t^{-m+1}/(m-1) + a\_{k+1} + a_k, \quad \text{(IV. 37)}$$

$\forall\ 0 < t \leq 1$, where we have that

$$(*)|u^{(k)}(t)| \leq 2t^{-m+1}(a_{k+1} + a_k), \quad \text{(IV. 38)}$$

$\forall\ 0 < t \leq 1$. Then using (*), we have

$$(**)|u^{(k)}(t)| \leq 2^{m-1}\, t^{-1}(a_k + \ldots + a_{k+m}), \quad \text{(IV. 39)}$$

$\forall\ 0 < t \leq 1$. Applying (**), to the case $k = 1$, we find that

$$|u^{(1)}(t)| \leq 2^m \log(1/t)(a_1 + \ldots + a_{m+1}), \qquad \text{(IV. 40)}$$

$$= u^{(1)}(1) - \int_t^1 2^{m-1}\, t^{-1}(a_k + \ldots + a_{k+m})dt, \qquad (k = 1),$$

to $0 < t \leq 1$.

If we integrate both members of the before inequality, we obtain the estimation that affirms the lemma to $u^{(0)} = u$. In effect, only we consider the integral

$$u^{(0)} = \int_0^t u^{(1)}(t)dt = \int_0^t 2^m \log(1/t)(a_1 + \ldots + a_{m+1})dt, \qquad \text{(IV. 41)}$$

and we arrive to that

$$|u^{(0)}(t)| = |u(t)| \leq C(a_0 + \ldots + a_{m+1}).$$

Now, the conclusion of the schollium with the identity

$$(***)\quad p^{\sim}\Phi_f^{T}(H) = \Phi_{pf}^{T}(H),$$

$\forall\ f \in \mathfrak{S}(\mathfrak{g})$ and $H \in \mathfrak{t}'$, will implies that if $X \in C$, then

$$|p\Phi_f^{T}(X)| \leq E\, |\pi_n(x_0)|^{-u} \Sigma^{u}{}_{q+1_{k=0}}\, \mu_{x_0(k)_p}(f), \qquad \text{(IV. 42)}$$

with E, a independent constant of f. Seeing that if $p(X) = -B(X, X)$, then $p(X) = ||X||^2$, $\forall\ X \in \mathfrak{t}$, which implies that

$$\Phi^{T}_{p^k f}(X) = ||X||^{2k} \Phi_f^{T}(X), \qquad \text{(IV. 43)}$$

Then due to that the integral $\int_G f(Ad(g)Y)dg$, converges absolutely $\forall\ Y \in \mathfrak{t}''$, and $f \in C^{\infty}(\mathfrak{g})$ (that defines a smooth function, $g(Y)$ $\forall\ g \in \mathfrak{t}''$), we have that the image $\Phi_f^{T} \in \mathfrak{S}(\mathfrak{t}'')$. Then the map

$$\mathfrak{S}(\mathfrak{g}) \rightarrow \mathfrak{S}(\mathfrak{t}''),$$

is continuous. ▪

Now we study the orbital integrals to other subalgebras of Cartan.

Let (P_0, A_0), be a minimal p-pair to G. Let $\mathfrak{h}$, be a subalgebra of Cartan of $\mathfrak{g}$. Then by the proposition that says that if H, is a Cartan subgroup of G, then there exists a standard cuspidal p-pair [Nolan Wallach], which implies that exist a canonical cuspidal p-pair (P_F, A_F), and $x \in G^0$, such that

$$\mathfrak{h} = Ad(x)\mathfrak{h}_F, \qquad \text{(IV. 44)}$$

Let H and H_F, be the corresponding subgroups of Cartan of G corresponding to the algebras $\mathfrak{h}$, and $\mathfrak{h}_F$.

For definition of Subalgebra of Cartan

$$\mathfrak{h}_F = \mathfrak{t}_F \oplus \mathfrak{a}_F, \qquad \text{(IV. 45)}$$

where $\mathfrak{t}_F$, is maximum Abelian in $\mathfrak{m} \cap \mathfrak{T}$.

Let T_F, be the subgroup of Cartan of ${}^0M_F$, corresponding to $\mathfrak{t}_F$. Then is easy to see that

$$H_F = T_F A_F = xHx^{-1}, \qquad \text{(IV. 46)}$$

$\forall\ x \in G^0$. Indeed, we consider the space

$${}^0M_F = \{m \in M_F \mid Ad(m) = I,\ \forall\ Ad \in End(M_F)\}, \qquad \text{(IV. 47)}$$

and let $T_F \subset {}^0M_F$, be the corresponding subgroup of Cartan of ${}^0M_F$, defined explicitly

$$T_F = \{m \in {}^0M_F \mid Ad(m)|_{\mathfrak{h}} = I\}, \qquad \text{(IV. 48)}$$

and such that T_F, $= \exp(\mathfrak{t}_F)$, with $\mathfrak{t}_F = \mathfrak{h}_F \cap \mathfrak{T}$. Then $\forall\ m \in {}^0M_F$, and $x \in G^0$,

$$Ad(m)|_{\mathfrak{h}}\, x = Ad(h)x = hx = xh = x,$$

since that furthermore ${}^0M_F$, is Abelian. Then $xh = x$, $\forall\ x \in G^0$, which is similar to

$$xhx^{-1} = 1_{\mathfrak{h}_F},$$

since that furthermore $\mathfrak{t}_F = \mathfrak{h}_F \cap \mathfrak{T}$, and H_F, is the subgroup of Cartan of H. Then $\forall\ h \in H$ and $x \in G^0$,

$$xHx^{-1} = H_F,$$

Then by the corresponding $\exp(\mathfrak{t}_H) = T_F$, and $\exp(\mathfrak{t}_F \oplus \mathfrak{a}_F) = T_F A_F = H_F = \exp(\mathfrak{h}_F)$, $\forall \mathfrak{t}_F = \mathfrak{h}_F \cap \mathfrak{T}$, we have finally

$$H_F = T_F A_F = xHx^{-1},$$

$\forall\ x \in G^0$. ■

On H_F, we have the invariant measure $dt_F da_F$, where dt_F, is the normalized invariant measure on T_F, and da_F, is the corresponding Lebesgue measure to an ortonormal base of $\mathfrak{a}_F$. Easy is to see that this measure is independent of the elections done in their definition.

We fix an invariant measure of G, and we have the quotient measure dgH on G/H. Let

$$\Phi^+ = \{\alpha \in \Phi(\mathfrak{g}_C, \mathfrak{h}_C) \mid Re\alpha > 0\}, \qquad \text{(IV. 49)}$$

with $\pi = \prod_{\alpha \in \Phi^+} \alpha$. Be $\Phi_R{}^+ = Re(\Phi^+)$. Let

$$\mathfrak{h}' = \{h\in\mathfrak{h} \mid \alpha(h) \neq 0, \forall\alpha\in\Phi\}, \qquad \text{(IV. 50)}$$

If $h\in\mathfrak{h}'$, then we define the correspondence

$$\varepsilon(h) = \mathrm{sgn}(\Pi_{\alpha\in\Phi^{R}+}\,\alpha(h)), \qquad \text{(IV. 51)}$$

If $f\in\mathfrak{S}(\mathfrak{g})$, we can to define the operator on $\mathfrak{h}'$,

$$\Phi_f^{H}(h) = \varepsilon(h)\,\pi(h)\int_{G/H} f(Ad(g)h)d(gH), \qquad \text{(IV. 52)}$$

with the domain of Φ_f^{H}, equal to the set of all the $h\in\mathfrak{h}'$, to which the integral converge absolutely.

Note that Φ_f^{H}, depends on the election of Φ^{+}, but only in sign. If in $\Phi(\mathfrak{g}_C, (\mathfrak{h}_F)_\sigma)$, we elect like space of positive roots to the set

$$\{\alpha \circ Ad(x)^{-1} \mid \alpha\in\Phi^{+}\}, \qquad \text{(IV. 53)}$$

then

$$\Phi_f^{H}(h) = \Phi_f^{H_F}(Ad(x)h), \qquad \text{(IV. 54)}$$

$\forall$ $x\in G$. Indeed, $\alpha \circ Ad(x)^{-1} = \alpha$, $\forall\alpha\in\Phi^{+}$,if only if, $\alpha = Ad(x)\alpha$, $\forall\alpha\in\Phi^{+}$, since $Ad(x)h = h$, $\forall$ $x\in G$. Then

$$\Phi_f^{H}(h) = \Phi_f^{Ad(H)|_{\mathfrak{h}}}(Ad(x)h) = \Phi_f^{H_F}(Ad(x)h),$$

$\forall$ $x\in G$, since $Ad(x)|_{\mathfrak{h}}\in H_F$, $\forall$ $x\in G$. We can consider the particular case when $H = H_F$, to these integral formulas.

We consider on G/A_F, the corresponding quotient measure to our election of invariant measures on A_F. Then is clear that

$$\Phi_f^{H}(h) = \varepsilon(h)\,\pi(h)\int_{G/A_F} f(Ad(g)h)d(gA_F), \qquad \text{(IV. 55)}$$

Now well, the lemma A. 1, implies that the invariant measures on K, ${}^0M_F$ and N_F, can be normalized such that

$$\Phi_f^{H}(h) = \varepsilon(h)\,\pi(h)\int_{K\times{}^0M\times N} f(Ad(knm)h)dkdmdn, \qquad \text{(IV. 56)}$$

Let $h\in\mathfrak{h}$. If $n\in N_F$, then we define the map

$$T_{\mathfrak{h}}: N_F\rightarrow Ad(N_F)\mathfrak{h}\setminus\mathfrak{h}, \qquad \text{(IV. 57)}$$

with rule of correspondence

$$n \mapsto Ad(n)\mathfrak{h} - \mathfrak{h}, \qquad \text{(IV. 58)}$$

If $n \in N_F$, then $n = \exp X$, $\forall X \in \mathfrak{n}_F$. If we develop the exponential series to

$$Ad(n) = e^{adX}, \tag{IV. 59}$$

Then $T_h(n) \in \mathfrak{n}_F$.

Indeed, considering that

$$\mathfrak{n}_F = \{X \in \mathfrak{n} \mid (adX - I)^k v = 0,\ \forall\ v \in F \text{ and } k \in \mathbf{Z}^+\}, \tag{IV. 60}$$

and

$$Ad(n) = \exp(adX) = \sum_{k=0}^{\infty} (1/k!)(adX)^k, \tag{IV. 61}$$

Then

$$T_h(n) = Ad(n)h\text{-}h = \sum_{k=0}^{\infty} (1/k!)(adX\text{-}I)^k h = \sum_{k=0}^{\infty} (1/k!)(adXh\text{-}Ih)^k = \Sigma_{k=0}^{\infty} T_h{}^k(n),$$

$\forall\ n \in N_F$. But $(adXh - Ih)^k = (adX - I)^k h = 0$, then $T_h{}^k(n) = 0$, $\forall\ k \in Z^+$. Thus $T_h(n) \in \mathfrak{n}_F$. An obvious evaluation give

$$(dT_h)_n(X) = Ad(n)[X, h], \tag{IV. 62}$$

In effect, for one side

$$dT_h(n)X = d(Ad(n)h)X = (dAd(n)X - X)h = [(adX - X)h]_n,$$

For other side

$$Ad(n)[X, h] = [Ad(n)X, Ad(n)h] = Ad(n)(Xh - hX) = Ad(n)Xh - Ad(n)hX,$$

But $Ad(n)X = X$, $\forall\ n \in N_F$, and $X \in \mathfrak{n}_F$. Then

$$Ad(n)Xh - Ad(n)hX = Ad(n)Xh - Xh = (Ad(n)X - X)h = dT_h(n)(X),$$

Thus (IV. 62), is verified $\forall\ n \in N_F$, and $X \in \mathfrak{n}_F$. This implies that

1. If $det(ad(h)|_{\mathfrak{n}}) \neq 0$ then T_h, is regular everywhere.

Note: $(adX)(h) = dT_h(X) \neq 0$, then T_h, is regular everywhere.

Lemma IV. 2. If $det(ad(h)|_{\mathfrak{n}}) \neq 0$ then T_h, is a diffeomorphism of N_F, in $\mathfrak{n}_F$, such that

$$det(ad(h)|_{\mathfrak{n}})\int_{N_F} f(Ad(n)h - h)dn = \int_{\mathfrak{n}_F} f(X)dX, \tag{IV. 63}$$

to f, a rapidly decreasing function on $\mathfrak{n}_F$.

If we demonstrate that T_h, is a diffeomorphism of N_F, in $\mathfrak{n}_F$, then the integration formula can follows of the formula

$$\Phi_f^H(h) = \varepsilon(h)\,\pi(h)\int_{G/N_F} f(Ad(g)h)d(gN_F), \tag{IV. 64}$$

to the differential of T_h. Let $h_0 \in \mathfrak{a}_F$, be such that $\alpha(h_0) > 0$, $\forall \alpha \in \Phi(P_F, A_F)$. Let $a_t = \exp(th_0)$, be then $T_h(a_t n a_{-t}) = Ad(a_t)T_h(n)$. Since T_h, is in particular, regular in 1_{N_F}, exist an open neighborhood U_0, of $0 = 1_{\mathfrak{n}F}$, in $\mathfrak{n}_F$, such that T_h, of U_1, in U_0, is a diffeomorphism.

In effect,

$$T_h(a_t n a_{-t}) = Ad(a_t n a_{-t})h$$

$$= Ad(a_t)Ad(n)Ad(a_{-t})$$

$$= Ad(a_t)Ad(a_{-t})Ad(n)h$$

$$= Ad(a_t)T_h(n),$$

Then

$$\bigcup_{t \geq 0} Ad(a_t)U_0 = \mathfrak{n}_F, \tag{IV. 65}$$

of which the equi-variance $T_h(a_t n a_{-t}) = Ad(a_t)T_h(n)$, implies that T_h, is suprajective, since

$$\bigcup_{t \geq 0} Ad(a_t)U_0 = T_h(a_t n a_{-t}) = \mathfrak{n}_F,$$

Now only is necessary to demonstrate the injectivity of T_h. To it, suppose that $T_h(n_1) = T_h(n_2)$. Let t, be such that $a_t n a_{-t} \in U_1$, $\forall$ j = 1, 2. Then

$$T_h(a_t n_1 a_{-t}) = Ad(a_t)T_h(n_1) = Ad(a_t)T_h(n_2) = T_h(a_t n_2 a_{-t}),$$

Thus $a_t n_1 a_{-t} = a_t n_2 a_{-t}$. Thus $n_1 = n_2$. Then T_h, is injective. Then T_h, is bijective. Thus T_h, is a diffeomorphism of N_F, in $\mathfrak{n}_F$.

Now we elect Φ^+, such that if $\alpha \in \Phi^+$, and if $\alpha|_{\mathfrak{a}} \neq 0$, then $\alpha|_{\mathfrak{a}} \in \Phi(P_F, A_F)$. Let Σ, be the set of all the $\alpha \in \Phi^+$, whose restriction to $\mathfrak{a}_F$, is not vanishing, to know;

$$\Sigma = \{\alpha \in \Phi^+ \mid \alpha|_{\mathfrak{a}F} \neq 0\}, \tag{IV. 66}$$

If $h \in \mathfrak{h}$, then $|\det(adh|_{\mathfrak{n}F})| = |\Pi_{\alpha\in\Sigma}\alpha(h)| = \varepsilon(h)\,\Pi_{\alpha\in\Sigma}\alpha(h)$.

This combined with the lemma IV. 2., implies that

2. $\Phi_f^H(h) = C_F\varepsilon(h)\,\Pi_{\alpha\in\Sigma}\alpha(h)\int_{K\times M\times N} f(Ad(k)Ad(m)(h + X))\,dk\,dm\,dn,$
If f is a soft function on $\mathfrak{g}$, then

$$f^{\sim}(X) = \int_K f(Ad(k)X)\, dk, \tag{IV. 67}$$

Since $Ad(^0M)$, preserve dX, on $\mathfrak{n}_F$, we have with the notation upward that:

3. $\Phi_f^H(h) = C_F \Pi_{\alpha\in\Phi^+ - \Sigma}\alpha(h)\int_{0M \times \mathfrak{n}} f^{\sim}(Ad(m)(h + X))\, dm\, dX,$

If $f\in\mathfrak{S}(\mathfrak{g})$, and $Q = P_F$, then to $Z\in\mathfrak{n}_F$,

4. $f^{(Q)}(Z) = \int_{\mathfrak{n}F} f(Z + X)\, dX,$

If $f\in\mathfrak{S}(\mathfrak{g})$, and if $h\in\mathfrak{h}$, then we write $h = h_{-} + \eta_{+}$, $\forall\, h_{-}\in\mathfrak{a}_F$, and $h_t\in\mathfrak{t}_F$, then

$$u(Z) = u(f, h_{-})(Z) = f^{\sim(Q)}(h_{-}tZ), \tag{IV. 68}$$

$\forall\, Z\in{}^0\mathfrak{n}_F$, of where is demonstrated that

$$\Phi_f^H(h) = C_F\Phi_u^T(h_+), \tag{IV. 69}$$

Then the evaluations in (1), (2), (3), (4), and (5), imply the following theorem of Harish-Chandra:

Theorem IV. 2.

1. The integral that defines Φ_f^H, $f\in\mathfrak{S}(\mathfrak{g})$, converges absolutely to $h\in\mathfrak{h}'$.
2. Let $\mathfrak{h}'' = \{h\in\mathfrak{h} | \alpha(h) \neq 0, \forall\alpha\in\Phi_{F,n}\}$. If $f\in\mathfrak{S}(\mathfrak{g})$, then Φ_f^H, is extended to be a smooth function on $\mathfrak{h}''$.

The theorem IV. 1., is followed of the behalf formulas.

If $X\in\mathfrak{g}$, then

$$\det(adX - tI) = \Sigma\, t^j D_j(X), \tag{IV. 70}$$

with dim $\mathfrak{g}$ = n. Let $D = D_r(X)$, with dim $\mathfrak{h}$ = r. Then from theorem of Harish-Chandra (theorem IV. 2), is followed:

Corollary IV. 1. $|D|^{1/2}$, is locally integrable on $\mathfrak{g}$.

Proof. Let the integral formula of Weyl on $\mathfrak{g}$,

$$\int_{\mathfrak{g}} f(X)dX = \Sigma C_j\int_{\mathfrak{h}j}|\pi_j(h)|^2\int_{G/Hj} f(Ad(\mathfrak{g})h)d(\mathfrak{g}H_j)dh_j \tag{IV. 71}$$

In the sense of that the right side converges if the left side that does. By appendix A,

$|D(h)|=|\pi_j(h)|^2$, $\forall\, h\in\mathfrak{h}_j$.

Let $f\in C^{\infty}{}_c(\mathfrak{g})$, be non-negative. Then by the theorem IV. 2., $\Phi_f\in S(\mathfrak{h}_j)$, $\forall\, j = 1, 2, \ldots, r$. Thus

$$\Sigma C_j \int_{\mathfrak{h}_j} |\Phi_f(h)| dh = \Sigma C_j \int_{\mathfrak{h}'} |\pi_j(h)| \int_{G/H_j} f(Ad(g)h) dgH_j \, dh_j$$

$$= \Sigma C_j \int_{\mathfrak{h}'} |\pi_j(h)| \int_{G/H_j} |\pi_j(h)| |D(Ad(g)h)|^{-1/2} f(Ad(g)h) dgH_j \, dh_j$$

$$= \Sigma C_j \int_{\mathfrak{h}'} |\pi_j(h)|^2 \int_{G/H_j} |D(Ad(g)h)|^{-1/2} f(Ad(g)h) dgH_j \, dh_j$$

$$= \int_{\mathfrak{g}} |D(X)|^{-1/2} f(X) dX.$$

■

Chapter V

Integral Transforms on Lie Groups and Lie Algebras

V. 1. Harish-Chandra function

Harish-Chandra construct a spherical function, that have the finality of shape an estimation criteria to the rapidly decreasing of the $\mathfrak{n}_F$–modules V˜, and the bound of the matrix coefficients under the asymptotic behavior that they apply. Such criteria come established under the range of definition of the function Ξ, of Harish-Chandra.

Let V a admissible ($\mathfrak{g}$, K)-modules. Let

$$V^{\sim} = \{\mu \in V^{*} | K\mu, \text{ generates a subspace of finite dimension of } V^{*}\}, \quad \text{(V. 1.1)}$$

V˜, is also an admissible ($\mathfrak{g}$, K)-module. Give a characterization of vector space of the admissible ($\mathfrak{g}$, K)-module V˜.

Lemma V. 1.1. Let W, a ($\mathfrak{g}$, K)-module. Suppose that exist a complex bilinear map

$$b : V \times W \rightarrow C, \quad \text{(V. 1.2)}$$

such that

1. $b(X, v, w) = -b(v, X, w)$, $b(kv, kw) = b(v, w)$ $\forall$ $v \in V$, $w \in W$, $X \in \mathfrak{g}$ and $k \in K$.
2. If $b(v, w) = 0$, then $v = 0$, and if $b(V, W) = 0$, then $w = 0$. Then W, is a ($\mathfrak{g}$, K)-isomorphic module with V˜.

Proof. The idea is demonstrate that exist a $\mathfrak{g}$, K-invariant between the subspaces W, and V˜, of the homomorphism spaces $\mathrm{Hom}_{\mathfrak{g},K}(V, W)$ and $Hom_{\mathfrak{g},K}(V^{*}, V^{\sim})$. We define

$$T: Hom_{\mathfrak{g},K}(V, W) \rightarrow Hom_{\mathfrak{g},K}(V^{*}, V^{\sim}), \quad \text{(V. 1.3)}$$

whose rule of explicit correspondence $\forall$ $w \in W$, is $T(w)(v) = b(v, w)$. Then, in principle T, is a belonging homomorphism to the space $Hom_{\mathfrak{g},K}(W, V^{\sim})$. Indeed, consider to the composition

$$T = T^{\sim} \circ T^{\wedge}, \quad \text{(V. 1.4)}$$

where $T^{\sim} \in Hom_{\mathfrak{g},K}(V, W)$, and $T^{\wedge} \in Hom_{\mathfrak{g},K}(V^{*}, V^{\sim})$. Since T˜, and T^, are homomorphism, these satisfies that $T^{\sim}(V) \subseteq W$, and $T^{\wedge}(V^{*}) \subseteq V^{\sim}$. Then

$$T(W) = T^{\sim} \circ T^{\wedge}(W) \subseteq V^{\sim}, \quad \text{(V. 1.5)}$$

Since $b(v, w) \neq 0$, $\forall\ v \in V$, $w \in W$, if and only if $\mathit{ker}T = \{0\}$, if and only if T, is injective. Let $\gamma \in K^{\wedge}$, and $V(\gamma)$, be their corresponding representation in the $(\mathfrak{g}, K)$-module V. We define to $\gamma^* = V^{\sim}(\gamma)$, such that $\forall\ T \in \mathit{Hom}_{\mathfrak{g}, K}(W, V^{\sim})$, $T(V(\gamma)) \subseteq W(\gamma^*)$. But by the non-degenerate of b, given that $V \neq 0$, and the K-invariance of b,

$$\mathit{dim}\ V(\gamma) = \dim\ (W(\gamma^*)), \tag{V. 1.6}$$

then $T(V(\gamma)) = W(\gamma^*)$. Thus T, is suprajective.

Then $T \in \mathit{Hom}_{\mathfrak{g}, K}(W, V^{\sim})$, is a $(\mathfrak{g}, K)$-isomorphism of W, a $V^{\sim}$. ■

Let $P = {}^0MAN$, be the minimal parabolic subgroup of G, with G, real and reductive. Let $(\pi_{\mu, \sigma}, H^{\mu, \sigma}) = \mathrm{Ind}_P{}^G(\sigma_\mu)$, be with σ_μ, the representation of P, given for

$$\sigma_\mu(man) = a^{\mu+\rho}\sigma(m), \tag{V. 1.7}$$

$\forall\ m \in {}^0M$, $a \in A$, and $n \in N$, $\mu, \rho \in (\mathfrak{a}_C)^*$. Let $\sigma^{\sim}$, be the dual representation of σ.

Lemma V. 1.2. $(H^{\mu, \sigma})_{K\sim} = (H^{\sigma\sim, -\mu})_K$.

Proof. Let $f \in (H^{\mu, \sigma})_K$, and $g \in (H^{\sigma\sim, -\mu})_K$, be then we realize

$$<f, g> = \int_K <f(k), g(k)>\, dk, \tag{V. 1.8}$$

We can verify the properties of the bilinear form as inner product in the $(\mathfrak{g}, K)$-module W (exercise).

Indeed, $\forall\ X \in \mathfrak{g}$, and considering to σ, a linear conjugate isomorphism, to know;

$$\sigma_\mu(f)(g) = <f, g>, \tag{V. 1.9}$$

that is to say, the isomorphism such that

$$(H^\infty)'_K = \sigma(H_K) = (H_K)^{\sim}, \tag{V. 1.10}$$

we have then that,

$$\sigma_\mu(f)(g) = <X\mu f, g>$$

$$= \int_K <X\mu f(k), g(k)>\, dk$$

$$= \int_K <f(k), -X\mu g(k)>\, dk$$

$$= -\int_K <f(k), X\mu g(k)>\, dk$$

$$= -<f, X\mu g> = \sigma_{-\mu}(g)(f) = \sigma^{\sim}{}_\mu(f)(g),$$

For other side (V. 1. 9), is non-degenerate, since $\sigma_{\mu}(H_K)(g) = \langle H_K, g\rangle = 0$, if $g = 0$, and $\sigma^{\sim}_{\mu}(f)(g) = \langle f, H_K\rangle = 0$, if $f = 0$. Thus σ_{μ}, is a isomorphism between the spaces $(H^{\mu,\sigma})_{K\sim}$, and $(H^{\sigma\sim, -\mu})_K$.

We define a functional Ξ_{μ}. Consider $\mu \in (\mathfrak{a}_C)^*$, $k \in K$ and $g \in G$. Then the functional Ξ_{μ}, that through of the representation $\pi_{\mu}(g)$, is realized in the class space $(H^{\mu})_K(\gamma) = \mathbf{C}I$, come defined as:

$$\Xi_{\mu}(g) = \langle \pi_{\mu}(g)\mathbf{1}_{\mu}, \mathbf{1}_{\mu}\rangle, \qquad \text{(V. 1.11)}$$

where is clear that

$$\pi_{\mu}(g)(\mathbf{1}_{\mu}) = \langle \pi_{\mu}(g)\mathbf{1}_{\mu}, \ \mathbf{1}_{\mu}\rangle = \int_K a(kg)^{\mu+\rho}\, dk,$$

$\forall\ g \in G$, where $\mathbf{1}_{\mu}(g) = a(g)^{\mu+\rho}$, and $\mathbf{1}_{\mu}(g) = 1$, $\forall\ g \in G$, $k \in K$. We call $\Xi = \Xi_0$. Then we define the function Ξ, of Harish-Chandra as follows:

Def. V. 1.1. Let $g \in {}^0MAN$. We define to the function $\Xi(a)$, like a solution to the integral equation

$$\int_K f(nak)a(k)^{\mu+\rho}\, dk = f(n)f(a), \qquad \text{(V. 1.12)}$$

$\forall$ n, $a \in {}^0MAN$, where explicitly whose function have a range defined by

$$C_t a^{-\rho}(1 + \log\|a\|^d) \geq \Xi(a) \geq a^{-\rho}, \qquad \text{(V. 1.13)}$$

Theorem V. 1.1. Exists positive constants C, and $d > 0$, such that

$$a^{-\rho} \leq \Xi(a) \leq C a^{-\rho}(1 + \log\|a\|^d), \qquad \text{(V. 1.14)}$$

$\forall\ a \in Cl(A^+)$.

Proof. Let (π, H), be denoted by (π_0, H^0), under the functional $\langle\, ,\, \rangle$, on $H_K = (H_K)^{\sim}$. To this only is required remember that we define the functional isomorphism

$$\sigma: H_K \times H_K \rightarrow (H_K)^{\sim}, \qquad \text{(V. 1.15)}$$

with rule of correspondence

$$(v, w) \mapsto \sigma(v)w = \langle v, w\rangle, \qquad \text{(V. 1.16)}$$

$\forall$ v, $w \in H_K$. Likewise, by the lemma V. 1. 2, extending said product to all the compact subgroup K, it goes over to that $\sigma(H_K) = (H_K^{\infty})' = (H_K)$. Thus $H_K = (H_K)^{\sim}$, under the functional $\sigma(.)(,) = \langle\, ,\, \rangle$.

Let 1_0, be the identity of the class space $H_K^{\infty}(\gamma_0) = \mathbf{C}I$. Let $1_0 \in H_K^{\infty}(\gamma_0)$. Then

$$\Xi(g) = <\pi(g)1_0, 1_0>, \qquad \text{(V. 1.17)}$$

Let V, be a ($\mathfrak{g}$, K)-submodule of H_K, generated by 1_0, that is to say

$$V = \Sigma\pi(K)1_0, \qquad \text{(V. 1.18)}$$

Then under < , >, $V^{\sim} = V$. Indeed,

$$\Xi(g) = <\pi(g)1_0, 1_0> = \int_K <\pi(g)1_0(k), 1_0(k)> dk$$

$$= \int_K <1_0(k), \pi(g)1_0(k)> dk$$

$$= <1_0, \pi(g)1_0>,$$

Thus $\Xi(g)[V] = V = V^{\sim}$. Suppose that $\mu + \rho$, is a weight of $\mathfrak{a}$, on $V/\mathfrak{n}V$, that is to say, $(V/\mathfrak{n}V)_{\mu+\rho}$, is a generalized eigenspace $\forall\mu, \rho\in(\mathfrak{a}c)^*$, and such that

$$\bigoplus_{\mu,\, \rho\in(\mathfrak{a}C)^*}(V/\mathfrak{n}V)_{\mu+\rho} = V/\mathfrak{n}V, \qquad \text{(V. 1.19)}$$

Let σ, be a ^{0}M-type of the weight space $(V/\mathfrak{n}V)_{\mu+\rho} = (H^{\sigma,\, \mu+\rho}) = H_\sigma{}^\mu$, with $\mathfrak{a}$, acting for $(\mu + \rho)I$. Then exist an element non-null in $Hom_{\mathfrak{g},\, K}(V, (H^\mu)_K)$. The reciprocity theorem of Frobenius implies that σ, can be of type ^{0}M. Then this implies that $\Xi_\mu = \Xi_0$. Then, due to that of the class space $(H_K{}^\mu)(\gamma_0) = \mathbf{C}I$, we can deduce directly of the isotopic class γ_0, that

$$\gamma_0(U(\mathfrak{g})^K) \subset U(\mathfrak{a})^W, \qquad \text{(V. 1.20)}$$

with $W = W(\mathfrak{g}, \mathfrak{a})$, if and only if $\Xi_\mu = \Xi_{s\mu}$, $\forall\ s\in W$, and $\mu\in(\mathfrak{a}c)^*$. In particular, if $\Xi_\mu = \Xi_0$, then $\mu = s = 0$, $\forall\ s\in W(G, A)$. Thus $\mu = 0$, where $\Lambda_V = \mu - \rho = -\rho$.

Then

$$\Xi(a) \leq Ca^{\Lambda V}(1 + \log\|a\|^d) \leq Ca^{-\rho}(1 + \log\|a\|^d), \qquad \text{(V. 1.21)}$$

Now we demonstrate the inferior inequality. By the formula

$$\Xi_\mu(a) = a^{\mu+\rho}\int_{\underline{N}} a(\underline{n})^{-\mu+\rho} a(a^{-1}\underline{n}a)^{\mu+\rho} d\underline{n}, \qquad \text{(V. 1.22)}$$

then

$$\Xi(a) = a^{\rho}\int_{\underline{N}} a(\underline{n})^{\rho} a(a^{-1}\underline{n}a)^{\rho} d\underline{n}, \qquad \text{(V. 1.23)}$$

Realizing the variable change described by the correspondence

$$\underline{n} \mapsto a\underline{n}a^{-1}, \qquad \text{(V. 1.24)}$$

we have

$$\Xi(a) = a^{-\rho}\int_{\underline{N}} a(a\underline{n}a^{-1}a(\underline{n})^{-1})^{\rho}a(\underline{n})^{2\rho}d\underline{n}, \qquad (V.\ 1.25)$$

Knowing that in G is defined a norm $\|\ldots\|$, exists μ, $\beta\in(\mathfrak{a}_C)^*$, with $\mu(H_j) > 0$, $\forall$ j, and C_1, $C_2 > 0$, (constants), such that

$$C_1a^{\mu} \leq \|a\| \leq C_2a^{\beta}, \qquad (V.\ 1.26)$$

$\forall$ $a\in Cl(A^+)$. Of this inequality is deduced immediately the inferior inequality. ■

Now we will demonstrate as Harish-Chandra has used the before result to prove the convergence of two important integrals.

Theorem V. 1.2. Let $d\in\mathbb{R}^+$. If $\varepsilon > 0$, and if $F\subset\Delta_0$, then

$$a^{-\rho}\int_{\underline{N}_F} a(\underline{n})^{\rho}(1 - \rho(\log a(\underline{n})))^{-d-\varepsilon}dn<\infty, \qquad (V.\ 1.27)$$

that is to say, said integral is convergent to a invariant measure on N_F.

Proof. Consider $h\in Cl(\mathfrak{a}^+)$, and a_t = *exp*th. Then the theorem V. 1. 1., implies that exist a positive constant such that

$$(a_t)^{\rho}\Xi(a_t) \leq C(1 + t)^{d}, \qquad (V.\ 1.28)$$

but by the theorem V. 1. 1., and the theorem D. 1., (Apendix D), is had that

$$\Xi(a)\leq (a_t)^{\rho}\Xi(a_t)= a^{-\rho}\int_{\underline{N}} a(\underline{n})^{\rho}a(a_t\underline{n}a_t^{-1})^{\rho}d\underline{n}, \qquad (V.\ 1.29)$$

We elect a $h\in Cl(\mathfrak{a}^+)$, such that

$$\begin{cases} \alpha(h) = 0, \forall\alpha\in F, \\ \alpha(h) = 1, \forall\alpha\in\Delta_0 - F, \end{cases} \qquad (V.\ 1.30)$$

then $\mathfrak{m}_F = C_{\mathfrak{g}}(h)$. Indeed, for other side

$$C_{\mathfrak{g}}(h) = \{h\in\mathfrak{a}^+ | Ad(g)h = h,\ \forall\ g\in G\}, \qquad (V.\ 1.31)$$

Then $\forall$ $g\in G$ and $h\in\mathfrak{a}$,

$$M = \{g\in G | Ad|_{\mathfrak{a}} = I\}, \qquad (V.\ 1.32)$$

But $M = \exp(\mathfrak{m}_C)$ then

$$\mathfrak{m} = \{h\in\mathfrak{a} | Ad(g)h = h, \text{ if and only if } [h, \mathfrak{g}] = 0\}, \qquad (V.\ 1.33)$$

Then $\mathfrak{m}_F = \mathfrak{a}_F \cap [\mathfrak{g}, \mathfrak{g}]$, of where

$$\mathfrak{m}_F = \{h\in\mathfrak{a}^+ | Ad(g)h = h,\ \forall\ g\in G\}, \qquad (V.\ 1.34)$$

Thus $\mathfrak{m}_F = C_g(h)$.

Let $^*\mathfrak{a}_F = \mathfrak{m}_F \cap \underline{\mathfrak{n}}_F$. Then $\underline{\mathfrak{n}}_F = {}^*\underline{\mathfrak{m}}_F \oplus \mathfrak{n}_F$. Due to a lemma on the normalization of all invariant measure on the subgroup $^*\underline{N}_F$, such that

$$\int_{\underline{N}F} f(^*\underline{\mathfrak{n}}_F)^2 d\underline{\mathfrak{n}}_F = 1, \qquad \text{(V. 1.35)}$$

By properties that are deduced of the normalization, can normalize the invariant measure on $\underline{N}_F$, such that if $f \in C_c(\underline{N})$, then

$$\int_{\underline{N}} f(\underline{n})d\underline{n} = \int_{^*\underline{N}F \times \underline{N}F} f(^*\underline{\mathfrak{n}}_F \underline{\mathfrak{n}}_F)\, d^*\underline{\mathfrak{n}}_F d^*\underline{\mathfrak{n}}_F, \qquad \text{(V. 1.36)}$$

of which we affirm is that

$$(a_t)^{\rho}\Xi(a_t) = \int_{NF} a(\underline{n})^{\rho} a(a_t \underline{n} a_t^{-1})^{\rho} d\underline{n}, \qquad \text{(V. 1. 37)}$$

Indeed, denoting by $\mathfrak{J}(t)$, the right side of (V. 1. 37), we have that if $a_t\, x a_{-t} = x$, $\forall\ x \in {}^*N_F$, and $x \in Na(x)k(x)$, is had that the transformation of (V. 1. 29), using (V. 1. 36), we give

$$\mathfrak{J}(t) = \int_{^*\underline{N}F \times \underline{N}F} a(^*\underline{\mathfrak{n}}_F)^{2\rho} a(k(^*\underline{\mathfrak{n}}_F)\underline{\mathfrak{n}}_F)^{\rho} a(k(^*\underline{\mathfrak{n}}_F) a_t \underline{n} a_t^{-1})^{\rho} d^*\underline{\mathfrak{n}}_F d^*\underline{\mathfrak{n}}_F, \qquad \text{(V. 1.38)}$$

Indeed, we do resource of that analytic expression of the function of Harish-Chandra

$$\Xi(a) = a^{-\rho} \int_{\underline{N}} a(n)^{2\rho} (a(a\underline{n}a^{-1}) a(\underline{n})^{-1})^{\rho} d\underline{n}, \qquad \text{(V. 1.39)}$$

which is a simple consequence of change of variable contemplated in (V. 1. 24), and effectuated in the function (V. 1. 23). If $a_t{}^*\underline{n}_F a_{-t} = {}^*\underline{n}_F$, $\forall\ {}^*\underline{n}_F \in {}^*N_F$, and $^*n_F \in N_F a(^*\underline{n}_F)k(^*\underline{n}_F)$, then

$$(a_t)^{\rho}\Xi(a_t) = \int_{\underline{N}} a(\underline{n})^{2\rho} (a(a\underline{n}a^{-1}) a(\underline{n})^{-1})^{\rho} d\underline{n}$$

$$= \int_{^*\underline{N}F \times \underline{N}F} a(a(a^*\underline{n}_F \underline{n}_F a^{-1}) a(^*n_t n)^{-1})^{\rho} a(^*\underline{n}_F \underline{n}_F)^{2\rho}\, d^*\underline{n}_F d\underline{n}_F$$

$$= \int_{^*\underline{N}F \times \underline{N}F} a(^*\underline{n}_F \underline{n}_F)^{-\rho} a(a^*\underline{n}_F \underline{n}_F a^{-1})^{\rho} a(^*\underline{n}_F \underline{n}_F)^{2\rho}\, d^*\underline{n}_F d\underline{n}_F,$$

But $k(^*\underline{n}_F) = \underline{n}_F$, since if $^*\underline{n}_F \in N_F a(^*\underline{n}_F)k(^*\underline{n}_F)$, and $k(^*\underline{N}_F) \subset K_F$, then

$$n_F = \underline{n}_F a(^*\underline{n}_F)k(^*\underline{n}_F), \qquad \text{(V. 1.40)}$$

then

$$(a_t)^{\rho}\Xi(a_t) = \mathfrak{J}(t),$$

$$= \int_{^*\underline{N}F \times \underline{N}F} a(^*\underline{n}_F)^{2\rho} a(k(^*\underline{n}_F)\underline{n}_F)^{\rho} a(k(\underline{n}_F) a_t \underline{n} a_{-t})^{\rho}\, d^*\underline{n}_F d\underline{n}_F,$$

having that $a(ka_t\underline{n}a_{-t}) = a(a_tk\underline{n}k^{-1}a_{-t})$, $\forall\ k\in K_F$, $t\in\mathbb{R}$, and $\underline{n}\in\underline{N}_F$. In K_F, compact $dk\underline{n}_Fk^{-1} = d\underline{n}_F$, on $\underline{N}_F$, $\forall\ k\in K_F$, then

$$\mathfrak{J}(t) = \int_{{}^*\underline{N}_F\times\underline{N}_F} a({}^*\underline{n}_F)^{2\rho}a(\underline{n}_F)^{\rho}a(a_t\underline{n}a_{-t})^{\rho}\, d^*\underline{n}_F d\underline{n}_F,$$

where the $\int_{{}^*\underline{N}_F} a({}^*\underline{n}_F)d^*\underline{n}_F = 1$, which will implies the integral expression of $(a_t)^{\rho}\Xi(a_t)$. In particular (V. 1.37), implies that

$$\int_{{}^*\underline{N}_F} a(\underline{n}_F)^{\rho}a(a_t\underline{n}_Fa_{-t})^{\rho}\, d\underline{n}_Fd^*\underline{n}_F \leq C(1 + t)^d, \qquad \text{(V. 1.41)}$$

If $F = \varnothing$, then to any $t \geq 0$,

$$a(a_t\underline{n}a_{-t})^{-2} = ||\sigma(a_t\underline{n}a_{-t})^{-1}v_0||, \qquad \text{(V. 1.42)}$$

is the polynomial in a subspace $W \subset\Delta_0$, to $X\in\underline{\mathfrak{n}}_F$. Developing $||\sigma(a_t\underline{n}_Fa_{-t})^{-1}v_0||^2$,

$$||\sigma(a_t\underline{n}_Fa_{-t})^{-1}v_0||^2 = ||v_0 + \Sigma_{j>0}e^{-jt}(\sigma(\underline{n}_F)^{-1}v_0)_j||^2 \leq$$

$$\leq 1 + ||e^{-jt}(\sigma(\underline{n}_F)^{-1}v_0)_j||^2$$

$$= 1 + e^{-jt}a(\underline{n}_F)^{-4\rho} \leq$$

$$\leq (1 + e^{-t}a(\underline{n}_F)^{-2\rho})^2,$$

where it has been demonstrated that

$$a(a_t\underline{n}_Fa_{-t}) \geq (1 + e^{-jt}a(\underline{n}_F)^{-2\rho})^{-1/2}, \qquad \text{(V. 1.43)}$$

$\forall\ t \geq 0$, and $\underline{n}_F\in\underline{N}_F$. If $r > 0$, then we define $(\underline{N}_F)_r = \{\underline{n}\in\underline{N}_F|\ a(\underline{n}) \geq r\}$. By a lemma on bounden subgroups of a real reductive group, is had that $(\underline{N}_F)_r$, is a bounded subgroup to any $r > 0$. For which said subgroup is compact $\forall\ r >0$. In (V. 1. 43), we take $t = -2\log r$, $\forall\ 0 < r < 1$. Then (V. 1. 41), implies that if $\underline{n}\in(\underline{N}_F)_r$, then

$$a(a_t\underline{n}_Fa_{-t}) \geq 2^{-1/2},$$

For which is necessary find that

$$C(1 + t)^d \geq \int_{(\underline{N}_F)_r} a(\underline{n})^{\rho}a(a_t\underline{n}_Fa_{-t})^{\rho}\, d\underline{n} \geq 2^{-1/2}\int_{(\underline{N}_F)_r} a(n)^{\rho}dn, \qquad \text{(V. 1.44)}$$

which implies

$$\int_{(\underline{N}_F)_r} a(\underline{n})^{\rho}d\underline{n} \leq C'(1 - 2\log r)^d, \qquad \text{(V. 1.45)}$$

Now we have $r_p = \exp(-2^p)$, $\forall\ p = 0, 1, \ldots$, with the notation of (*X), is had that

$$\int_{(\underline{N}_F)_r} a(\underline{n})^{\rho}d\underline{n} \leq C'(1 + 2^{p+1})^d \leq C(1/2)^{pd}, \qquad \text{(V. 1.46)}$$

If $\underline{n} \in (\underline{N}_F)_{r_{p+1}} - (\underline{N}_F)_r$, then $r_p \geq a(\underline{n})^\rho \geq r_{p+1}$. For which, on this same inequality we have

$$1 + 2^p \leq (1 - \rho(\log a(\underline{n})) \leq 1 + 2^{p+1}, \qquad \text{(V. 1.47)}$$

This implies that if $\varepsilon > 0$, then

$$\int_{(\underline{N}F)r_{p+1} - (\underline{N}F)r} a(\underline{n})^\rho (1 - \rho(\log a(\underline{n}))^{-d-\varepsilon} d\underline{n} \leq C'(1 + 2^p)^{-d-\varepsilon} 2^{pd} \leq C(1/2)^{pd} \leq C^m 2^{-\varepsilon p}, \qquad \text{(V. 1.48)}$$

If we add on $p > 0$, then

$$\int_{(\underline{N}F) - (\underline{N}F)r0} a(\underline{n})^\rho (1 - \rho(\log a(\underline{n}))^{-d-\varepsilon} d\underline{n} \leq C^m \sum 2^{-\varepsilon p} < \infty,$$

This implies that

$$\int_{\underline{N}F} a(n)^\rho (1 - \rho(\log a(\underline{n}))^{-d-\varepsilon} dn < \infty,$$

with $(\underline{N}_F)_r$, compact. ■

Theorem V. 1.3. If $r > 0$, and if $q > d + r$, then

$$\int_{\underline{N}F} a(n)^{\rho_F} \Xi_F(\underline{n})(1 + \rho(\log\|m_F(\underline{n})\|)^d (1 - \rho(\log a(\underline{n}))^{-q} d\underline{n} < \infty, \qquad \text{(V. 1.49)}$$

Proof. To prove such affirmation, first is necessary to prove that exist a constant $C > 0$, such that

$$1 + \log\|m_F(\underline{n})\| \leq C(1 + \log\|n\| - \rho \log a(\underline{n})), \qquad \text{(V. 1.50)}$$

$\forall \underline{n} \in \underline{N}_F$.

We assume:

i. $\|g\| = \|\sigma(g)\|$, with (σ, F), a representation of finite dimension of G, and $\|\sigma(g)\|$, a Hilbertian norm of $\sigma(g)$, relative to $<, >$, of F, such that

$$\sigma(g)^* = \sigma(\theta(g))^{-1},$$

ii. We elect a base $\{u_j\} \subset F$, such that the block of the representation of M_F, let be a diagonal block:

$$\begin{bmatrix} A_1\ldots\ldots & \ldots & \ldots & 0 \\ \ldots A_2\ldots\ldots & \ldots & \ldots & \ldots \\ \ldots\ldots\ldots\ldots & \ldots & \ldots & \ldots \\ 0\ldots\ldots\ldots & \ldots & \ldots & A_d \end{bmatrix},$$

and the elements of N_F, conform the block of the superior triangular form:

$$\begin{bmatrix} I_1\ldots\ldots & \ldots & \ldots & * \\ \ldots I_2\ldots\ldots & \ldots & \ldots & \ldots \\ \ldots\ldots\ldots\ldots & \ldots & \ldots & \ldots \\ 0\ldots\ldots\ldots & \ldots & \ldots & I_d \end{bmatrix},$$

then

$$\|\sigma(\underline{n})\sigma(\underline{n})^*\| = \|\sigma(n_F(\underline{n})\sigma(m_F(\underline{n}))\sigma(m_F(\underline{n}))^*\sigma(a_F(\underline{n}))^2\sigma(n_F(\underline{n}))^*\|,$$

where

$$\|\sigma(m_F(\underline{n}))\sigma(m_F(\underline{n}))^*\sigma(a_F(\underline{n}))^2\|\leq\|\sigma(\underline{n})\sigma(\underline{n})^*\|,$$

which implies that

$$\|\sigma(m_F(\underline{n}))\sigma(m_F(\underline{n}))^*\|\leq\|\sigma(a_F(\underline{n}))\|^{-2}\|\underline{n}\|^2, \qquad \text{(V. 1.51)}$$

If we apply the referent lemmas to the properties of boundess of a subgroup of a real reductive group, the consideration of the logarithm of the last inequality must be put yet, we have (V. 1. 50), $\forall \underline{n}\in\underline{N}_F$.

Now well, this last inequality, to the light of the properties of invariant integration on subgroups of a real reductive group, implies that is sufficient to demonstrate that the following integral is finite, $\forall\ q > d + r$, that is to say;

$$\mathfrak{I}(t) = \int_{\underline{N}F} a(n)^{\rho F}\Xi_F(\underline{n})(1 - \rho(\mathit{log}\,a(\underline{n}))^{r-q}\, d\underline{n}<\infty, \qquad \text{(V. 1.52)}$$

$\forall\ q > d + r$. But for definition of the function Ξ_F, actually, the function Ξ, to ${}^0M_F$, we have that said integral take the form

$$\mathfrak{I}(t) = \int_{\underline{N}F} a(n)^{\rho F}\int_{KF} a(km_F(\underline{n}))^{\rho}\, dk(1 - \rho(\mathit{log}\,a(\underline{n}))^{r-q}\, d\underline{n}, \qquad \text{(V. 1.53)}$$

If $k\in K_F$, and $\underline{n}\in\underline{N}_F$, then

$$k\underline{n} = knm_F(\underline{n})a_F(\underline{n})k_F(\underline{n}) = knk^{-1}km_F(\underline{n})k^{-1}a_F(\underline{n})kk_F(\underline{n}), \qquad \text{(V. 1.54)}$$

with $n\in N_F$. Thus $km_F(\underline{n}) = n_F(k\underline{n}k^{-1})$, and $a_F(k\underline{n}) = a_F(\underline{n})$. This will implies that

$$\mathfrak{I}(t) = \int_{\underline{N}F\times KF} a(k\underline{n}k^{-1})^{\rho F} a(m_F(k\underline{n}k^{-1}))(1 - \rho(\mathit{log}\,a(\underline{n}))^{r-q}\, d\underline{n}dk$$

$$= \int_{\underline{N}F\times KF} a(\underline{n})^{\rho F}(1 - \rho(\mathit{log}\,a(k\underline{n}k^{-1}))^{r-q}\, d\underline{n}dk \leq$$

$$\leq\int_{\underline{N}F} a(\underline{n})^{\rho F}(1 - \rho(\mathit{log}\,a(\underline{n}))^{r-q}\, d\underline{n}dk <\infty,$$

which is obvious by the theorem that establishes the existence of positive constant to continuous semi-norm (appendix C), in the study of asymptotic behavior of the matrix coefficients.■

V. 2. Twistor transform to ladder representations

Let $\mathbb{P}^+$, and $\mathbb{P}^+$, be open orbits of SU(2, 2), on $\mathbf{M} = Gr_2(\mathbf{C}^4)$, and $\mathbb{P} = \mathbb{CP}_3 = \mathbb{P}(\mathbb{C}^4)$, respectively. The dual projective twistor space is $\mathbb{P}^* = \mathbb{P}((\mathbb{C}^4)^*)$. The sheaves of holomorphic functions and of holomorphic 3-forms (that is to say, belonging to space $\Lambda^3(S^3)$))), are denoted by $\mathcal{O}$, and Ω^3, respectively.

The Penrose transform to this case result completely natural and, in particular this isomorphism with cohomology result be SU(2, 2)-equivariant.

Is clear that twistor description of this representation is one of the most clear than the version of Maxwell fields to effects of induction of G-modules. The action of SU(2, 2), is automatic. Now well, is natural question how the scalar product of the Hermitian form arises in this twistor correspondence.

Indeed, motivated by the classical construction is well acquaintance that this can be feasible by an integral intertwining operator. The cohomology $H^1(|\mathbb{P}^+|, \Omega^3)$, (where$|\mathbb{P}^+|$, represent the closure of the open orbit $\mathbb{P}^+$) give an ascent directly to fields on $\mathbb{P}^+$, considering that the description "potential module gauge", of the same fields finds their interpretation on the dual twistor space like $H^1(|\mathbb{P}^{*-}|, \mathcal{O})$.

Note that $|\mathbb{P}^+|$, is simply connect (contractible really), and have base neighborhood of Stain such that the original description of field and the description "potential modulo gauge", meet.

Thus, the integral intertwining operator, which establish the crucial ingredient in the classical construction is interpreted in the twistor description like a integral intertwining twistor operator, being this the twistor transform

$$\mathcal{T}: H^1(|\mathbb{P}^+|, \Omega^3) \rightarrow H^1(|\mathbb{P}^{*-}|, \mathcal{O}), \qquad \text{(V. 2.1)}$$

Considering this operator is easy describe the scalar product in a path in the which is necessary the SU(2, 2)-invariance. If $\Gamma \in H^1(|\mathbb{P}^{*-}|, \mathcal{O})$, then $\underline{\Gamma} \in H^1(|\mathbb{P}^-|, \mathcal{O})$, and to obtain the invariance under conformal transforms of such product, is necessary the multi-linear pairing of G-modules $H^1(|\mathbb{P}^+|, \Omega^3)$, and $H^1(|\mathbb{P}^-|, \mathcal{O})$,given by

$$H^1(|\mathbb{P}^+|, \Omega^3) \otimes H^1(|\mathbb{P}^-|, \mathcal{O}) \rightarrow \mathbb{C}. \qquad \text{(V. 2.2)}$$

Such pairing can be obtained taking representatives forms σ, on$|\mathbb{P}^{+}|$,of type (3, 1), and τ, on $|\mathbb{P}^{-}|$,of type (0, 1), and integrating the 5-form σ – τ, on$\mathbb{P}^{0}$. This is equivalent in the language of the homomorphism to

$$H^1(|\mathbb{P}^{+}|, \Omega^3) \otimes H^1(|\mathbb{P}^{-}|, \mathcal{O}) \to H^2(\mathbb{P}^{0}, \Omega^3) \to H^3(\mathbb{P}, \Omega^3) \to \mathbb{C}, \qquad \text{(V. 2.3)}$$

and by duality in twistor theory, can be deduced the scalar product wanted. Thus $\forall F$, $G \in H^1(|\mathbb{P}^{+}|, \Omega^3)$, we define

$$<F, G> = F.\underline{\mathcal{T}G}, \qquad \text{(V. 2.4)}$$

As exercise, consulting [Eastwood and Ginsberg, 1981, 177-196], is possible demonstrate the coincidence of this twistor construction with the classical construction on the massless fields.

The twistor construction has much advantage on the classical construction, being of major relevance the possibility of to be generalized to an ambit of spaces of major dimension, or of structures that are defined under topologies more weakly. Many important representations of reductive Lie groups occurring naturally as a cohomology on a homogeneous space, such is the case, for example, of some induced representations on generalized orbital spaces. In this point, one can have the possibility of to know these representations are unitarizables. To this goal, result useful the work with cohomology, eluding of this manage arguments of the space-time that can hinder the study of the unitarization of the representations.

Nevertheless, we have not that lose of view that we want obtain G-modules that can be classified like £-modules with the goal of obtain all the unitary representations of the integral operators that acts in the general solution of the Maxwell equations.

Thus want understand and generalize so much as be possible the twistor transform $\mathcal{T}: H^1(|\mathbb{P}^{+}|, \Omega^3) \to H^1(|\mathbb{P}^{*-}|, \mathcal{O})$, like integral operator in electrodynamics of type intertwining to the obtaining of solutions in the space-time modulo the Maxwell fields into of the context of the unitary representations.

To can use the formulation of the twistor transform like a ensemble of classical intertwining integral operators and give their corresponding representations is necessary to check that the Hermitian form originated of the twistor transform let be symmetric. Finally one can verify this on the L-types. This have been did by example to the ladder representations of SU(p, q).

Consider a vector complex space T, of dimension (N + 1), with Hermitian form Φ, with signature (p, q), p + q = N + 1. Consider by simplicity that $2 \leq p \leq q$ (if p = 1 (or pair if p = 0), the general process is valid but the conclusions are subject to certain little modifications [Eastwood, 1983]). Be G = SU(p, q), the subgroup of SL(N + 1, $\mathbb{C}$) the which preserve Φ. The projective space $\mathbb{P}^N(\mathbb{C})$, is sliced in, $\mathbb{P}^{+}$, $\mathbb{P}^{-}$y $\mathbb{P}^{0}$, under the action of G. Here $\mathbb{P}^{+}$(respectively $\mathbb{P}^{-}$), is the space of lines $x \subset \mathbf{T}$, such that $\Phi|_x$, is positive (respectively negative) defined.

The space $\mathbb{P}^0$,sometimes denoted by *PN*, consists of those lines on which Φ, is null. Thus $\mathbb{P}^0$, is a real hyper-surface in $\mathbb{P}^N(\mathbb{C})$. Note that $\mathbb{P}^+$, and $\mathbb{P}^-$, are indicating open subsets like complex manifolds. The fact of that these spaces are the unique open G-orbits do the construction of the appearing simpler.

Let **M**, be the p-dimensional Grassmanian subspaces of **T**. These open and *min*(p, q) + 1, G-orbits in **M**, corresponds to the possible ways in the which Φ, can be reduced like a non-degenerated form on a space of dimension p. We are interested in the orbit $\mathbf{M}^+$, on whose points Φ is reduced to be positive defined. Note that this orbit $\mathbf{M}^+$, is a Stein manifold.

As it is demonstrated originally in [Eastwood, 1983] the composition of two Penrose transforms establishes the isomorphism given by the twistor transform to SU(p, q), to know

$$\mathcal{T}: H^{p-1}(\mathbb{P}^+, \mathcal{O}(-n-p)) \rightarrow H^{q-1}(\mathbb{P}^{*-}, \mathcal{O}(n-p)), \qquad \text{(V. 2.5)}$$

The sheaf O(-k), correspond to the kth-power of the tautological bundle of lines on $\mathbb{P}^N(\mathbb{C})$. Thus when p = n = 2, $O(-4) = \Omega^3$, this meet with the map (V. 2. 1). The demonstration of that (V. 2. 1), and (V. 2. 5), are isomorphisms. For example, when $n \geq 1$, one demonstrate that the left side is isomorphic to fields on $\mathbf{M}^+$, satisfying the differential equations analogous to the massless equations of helicity *n*/2, on the Minkowski space. Similarly, the right side is naturally isomorphic to potentials module gauge to the same fields on $\mathbf{M}^{*-}$. But $\mathbf{M}^{*-}$, is the same like in $\mathbf{M}^+$, in the isomorphism. Newly one can to extend the isomorphism to the closures of $\mathbb{P}^+$, and $\mathbb{P}^-$, given the theorem:

Theorem V. 2.1. Exist a map SU(p, q)-equivariant

$$\mathcal{T}: H^{p-1}(|\mathbb{P}^+|, O(-n-p)) \rightarrow H^{q-1}(|\mathbb{P}^{*-}|, O(n-q)), \qquad \text{(V. 2.6)}$$

which is isomorphism.

Proof. [10].

V. 3. Radon-Schmid transform to D-modules like induced representations

We consider the Penrose transformation, [4] through of the correspondence

$$P \xleftarrow{\ \nu\ } F \xrightarrow{\ \pi\ } M \qquad \text{(V. 3.1)}$$

where $\mathbb{F} = \mathbb{F}_{1,2}(V)$, is the manifold of flags of dimension one and two, associated to 4-dimensional complex vector space V. Be $\mathbb{P} = \mathbb{F}_1(V)$, such that $\mathbb{F}_1(V) \cong \mathbb{P}^3(\mathbb{C})$, (complex lines in $\mathbb{C}^4$), and be $\mathbb{M} = \mathbb{F}_2(V)$, such that $\mathbb{F}_2(V) \cong G_{2,4}(\mathbb{C})$, (Grassmannian manifold of 2-dimensional complex subspaces), with $\mathbb{M} \cong \mathbb{R}^4 \otimes_{\mathbb{R}} \mathbb{C}$, where

$$\mathbb{M} = \{\underline{z} \in \mathbb{C}^4 \mid \underline{z} = (z_1, z_2, z_3, z_4), \forall\, z_i = x_i + jy_i, \forall\, x_i, y_i \in \mathbb{R}\}, \qquad (V.3.2)$$

is the 4-dimensional complex compactified Minkowski space [4]. The projections of $\mathbb{F}$, are given for:

$$\nu(L_1, L_2) = L_1, \qquad (V.3.3)$$

and

$$\pi(L_1, L_2) = L_2, \qquad (V.3.4)$$

where $L_1 \subset L_2 \subset V$, are complex subspaces of dimension one and two, respectively, defined a element (L_1, L_2), of $\mathbb{F}$, to know

$$\mathbb{F} = \{(L_1, L_2) \in V \times V \mid L_1 \subset L_2 \subset V,\ \nu(L_1, L_2) = L_1,\ \pi(L_1, L_2) = L_2\}, \qquad (V.3.5)$$

If $\mathbb{M}$, is a compactified Minkowski space [4] then

$$\{\textit{set of equations of massless fields}\} \cong \{dF = 0,\ dF^* = j,\ W \circ \delta = 0,\ R^{ij} = 0,\ R^{ij} - g^{ij}R = 0, \ldots\},$$

$$[UPVM] \qquad (V.3.6)$$

that is to say, is required the *spectral resolution of complex sheaves* [17], of seated certain class in the *projective space* $\mathbb{P}$, to give solution to the field equations modulus a flat conformally connection [10]

$$O^0_{\mathbb{P}}(h) \rightarrow \ldots \rightarrow O^i_{\mathbb{P}}(h) \rightarrow O^{i+1}_{\mathbb{P}}(h) \rightarrow \ldots \rightarrow 0, \qquad (V.3.7)$$

Let $\mathcal{P}$, be the *Penrose transform* [29] *associated to the double fibration* in (1), used to represent the holomorphic solutions of the *generalized wave equation* [4],with parameter of *helicity* h [29]:

$$\square_h \phi = 0, \qquad (V.3.8)$$

on some open subsets $U \subset \mathbb{M}$, in terms of *cohomological classes of bundles of lines* [4], on $\underline{U} = \nu(\pi^{-1}(U)) \subset \mathbb{P}'(\mathbb{P}'$, is the *super-projective space*).

Is necessary to mention that *these cohomological classes are the conformal classes* that are wanted determine to solve *the phenomenology of the space-time* to diverse interactions studied in *gauge theory* [19], and can construct a general solution of the general cohomological problem of the *space-time.*

With major precision we sign that bundle of lines on $\mathbb{P}$, are given $\forall$ $k \in \mathbb{Z}$, by the kth-tensor power $O_{\mathbb{P}}(k)$, of the tautological bundle [19], (is the bundle that serve to explain in the context of the bundles of lines on $\mathbb{P}$, the general bundle of lines of $\mathbb{M}$).

Let $h(k) = -(1 + k/2)$, be $\forall$ $x \in \mathbb{P}$, and let$\underline{x} = \pi(\nu^{-1}(x))$. Then a result that establish the *equivalences* on the cohomological classes of the bundle of lines on $\underline{U}$, and the family of solutions of the equations (V. 3. 8), (equations of the massless fields family on the Minkowski space $\mathbb{M}$, with helicity h) is the given by:

Theorem V. 3. 1.*(with classic Penrose transform)* Be $U \subset \mathbb{M}$, an open subset such that $U \cap \underline{x}$, is connect and simply connect $\forall$ $x \in \underline{U}$. Then $\forall$ $k < 0$, *the associated morphism to the twistor correspondence* (1); the which map a 1-form on $\underline{U}$, to the integral to along of the fibers ofπ, of their inverse image forν, induce an isomorphism of cohomological classes:

$$H^1(\underline{U}, O_{\mathbb{P}}(k)) \cong \ker(U, \square_{h(k)}), \tag{V. 3.9}$$

Part of the object of our research is centred in *the extension of the space of equivalences of the type* (V. 3. 8), under a more general context given through of the language of the *D-modules*, that is to say, we want *extend our classification of differential operators of the field equations* to the context of the homogeneous bundle of lines and obtain a complete classification of all the differential operators on *the curved analogous of the Minkowski space of* $\mathbb{M}$. Thus our moduli space will be those of *the equivalences of the conformal classes given in* (V. 3. 8), in the language of the *algebraic objects D-modules with coefficients in a coherent sheaf.*

We consider a correspondence

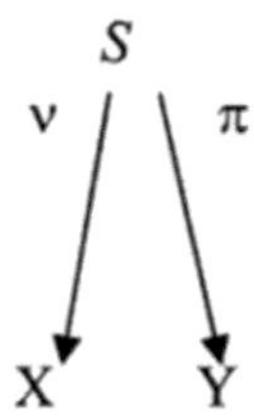

(V. 3.10)

where all the manifolds are analytic and complex, ν, and π, are proper and (ν, π), induces a closed embedding $S \hookrightarrow X \times Y$ [22]. Be $d_S = \dim_{\mathbb{C}} S$, with $d_{S/Y} = d_S - d_Y$.

We define the transform of a sheaf F on X, (more generally, of a object of the derived category of sheaves) like

$$\Phi_S F = R\pi \nu^{-1} F[d_{S/Y}], \tag{V. 3.11}$$

and we define the transform of a D_X-module $\mathcal{M}$, like $\underline{\Phi}_S\mathcal{M} = \underline{\pi}_* D v^* \mathcal{M}$, where $\underline{\pi}_*$, and v^*, denotes the direct and inverse images of π, and$v$, respectively, in the sense of the D-modules[1] [22], and we consider also

$$\Phi_{\underline{S}} G = R v_! \pi^{-1} G[d_{S/X}], \qquad (V.\ 3.12)$$

To a sheaf $\mathcal{G}$, on Y. Then is had the formula

$$\Phi_S R\, \mathrm{Hom}_{D_X}(\mathcal{M}, O_X) = R\, \mathrm{Hom}_{D_Y}(\underline{\Phi}_S\mathcal{M}, O_Y), \qquad (V.\ 3.13)$$

of the which it is deduce the formula, to $\mathcal{G}$, the sheaf on Y, (coherent sheaf):

$$R\Gamma(X; R\mathrm{Hom}_{D_X}(\mathcal{M} \otimes \Phi_{\underline{S}} \mathcal{G}, O_X)[d_X] \cong R\Gamma(Y; R\mathrm{Hom}_{D_Y}(\underline{\Phi}_S\mathcal{M} \otimes \mathcal{G}, O_Y)[d_Y], \qquad (V.\ 3.\ 14)$$

This define a categoric equivalence of the transformation (V. 3. 1), in the context of the *right derived D-modules*, $D^b(\mathcal{M}^R_{qc}(D^V))$, because is necessary to give a equivalence with a sub-category of the *right* D-modules that have support in Y, to of that form guarantee the inverse image of $\mathcal{P}$, and with it to obtain a image of closed range of the operator $\mathcal{P}$, conformed their uniquely [21, 22, 26].

***Theorem of Kashiwara* V. 3.2.** Let i : Y ◎ X, be of the correspondence (V. 3. 10), a closed immersion. Then the direct image functor i_+, is an equivalence of $M^R_{qc}(D^i)$, with full sub-category of $M^R_{qc}(D)$, consisting of modules with support in Y.

Proof: [46].

This equivalence preserves *coherency* and *holonomicity* [17, 19]. Then preserve *conformability* in $\mathbb{M}$[19].

Let's formulate in the language of the D-Modules and its sheaves, like was given in a resolution (V. 3. 7); the correspondence between the space of coherent D-Modules and the massless field equations space of the Penrose transform.

---

[1] To define the images of direct functors to D-modules is necessary use *derived categories*. For it, is simpler defining them for *right D-modules*. Let $D^b[M^R_{qc}(D^\bullet)]$, be the derived category bounded for right quasi-coherent D^V-modules of the form

$$Rv_+(\mathcal{V}) = Rv_\bullet(\mathcal{V} \otimes^R_{DV} D_{Y\to X}),$$

where $\mathcal{V}$,is the *characteristic manifold* and R, is the *right derived functor* following:

$$R : \mathcal{M}^R_{qc}(D^V) \to \mathcal{M}_{qc}(D).$$

Also r : Y → X, and $D_{Y\to X} = v^*(D) = O_Y \otimes_{v^{-1}(O_X)} v^{-1}D$. Then $D_{Y\to X}$, is a right $v^{-1}D$- module to the right multiplication in the second functor.

But for it, is necessary to include a result that establishes the regularity in the analytical sense of the Riemannian manifold, which shapes the space - time, and that allows the application of the involutive distribution theorem on integral submanifolds as solutions of the corresponding massless field equations on $G_{2,4}(\mathbb{C})$.

By the *Kashiwara* (*theorem V. 3. 1*), and some results of Oshima on involutive manifolds [46], we can characterize certain spaces to the regularity of the images of $\mathcal{P}$, in *D*-modules. These spaces are the induced representations that we want exhibite, such as Schmid demonstrate it [47].

V. 4. Penrose transform to locally compact representations

The roll of the Borel sub-algebras and the corresponding Cartan subalgebras are equivalents in different context, the first uses in complex analytic manifolds to characterize the points of a coherent sheaf of differential operators of the corresponding flag manifolds of a holomorphic complex bundle (bundle of Borel subalgebras), with cohomology group of dimension cero (theorem of Borel-Weil). The global sections in this coherent sheaf denotated for

$$\Gamma(X, \mathfrak{o}(\lambda)) = H^0(X, \mathfrak{o}(\lambda)), \qquad (V. 4.1)$$

conform an irreducible G-module of finite dimension with maximal weight λ, and Flag manifolds X, (which is a projective manifold).

For other side, the Cartan subalgebras are the unique global constant sections of the holomorphic complex bundle modulus their nilpotent radical bundle of Borel subalgebras, which are points of a Lie algebra of the general linear group subjacent in the reductive Lie group G. In others words the holomorphic bundle modulus a radical bundle result be a trivial vector fibred bundle on X, and is a subbundle of the bundle of Borel subalgebras, thus result natural to think that using certain structure more fine of the flag manifolds like for example, the given for open orbits of flag manifold and the continuous homomorphism between said open orbits can obtain an extension of the classification of the irreducible representations that exist in the space $H^0(X, \mathfrak{o}(\lambda))$, and that under the association of irreducible minimal K-types is obtained a class more wide of irreducible unitary representations classified for the theory of Langlands.

One of the important theorems to the obtaining of induced representations (with canonical P-pairs (P, A), with Levi decomposition $P = MN = {}^0MAN$, $P = M\underline{N}$, and $N = \theta(N)$, ($\underline{P} = \theta(P)$)), like ($\mathfrak{m}$, K_M)-subquotients of $V/\mathfrak{n}V$, is the theorem of Casselman, the which can be generalized to quotients of the form

$$V_{\mathfrak{n}0} = V_{\mathfrak{n}}/((V_{\mathfrak{n}})(^*\mathfrak{n})), \quad (V.\ 4.2)$$

on Frèchet spaces. Then in the structure of ($\mathfrak{g}$, K)-module of V, there is a structure of ($\mathfrak{g}$, K)-module of V, there is a structure of subquotient ($\mathfrak{m}_0$, K ∩ M)-module of $V_{\mathfrak{m}0}$, where in turn, it there is an analytic structure (C^{∞}), of Frèchet space of K-finite vectors such that V, is finitely generated and admissible like ($\mathfrak{g}$, K)-module. Then exist a continuous homomorphism between V, and the Langlands subrepresentation of data (P, σ, ν), belonging to space

$$Hom_{\mathfrak{g},\, K}(V, I_{P_0,\ \sigma,\ \nu - \rho^0}) \neq 0, \quad (V.\ 4.3)$$

Using a cohomology with respect to the representations of principal series we can characterize an isomorphism between ${}^0M$-modules like the obtained for the Casselman subrepresentations[2] and given for the cohomology group

$$H^j(\mathfrak{n}_P, E_{\lambda}) = \bigoplus_{s \in WP,\ l(s) = j} E_{s(\rho + \lambda) - \rho}, \quad (V.\ 4.4)$$

Where the weights $s(\rho + \lambda) - \rho$, are all distinct and dominating. Given that s, changes in all the range of W^P, then the decomposition of $H^*(\mathfrak{n}, F_{\lambda})$, like M_C-module, is of free multiplicity.

If we consider a nilpotent radical Lie algebra $\mathfrak{u}$, of a θ-stable parabolic subalgebra $\mathfrak{q}$, such that also to a subalgebra

$$\mathfrak{l} = \{X \in \mathfrak{g} \mid [X, H] = 0\}, \quad (V.\ 4.5)$$

it is satisfy that

$$\mathfrak{q} = \mathfrak{l} \oplus \mathfrak{u}, \quad (V.\ 4.6)$$

with

$$\mathfrak{u} = \bigoplus_{\alpha \in \Delta(\mathfrak{h},\, \mathfrak{g}),\ \alpha(ih) < 0} \mathfrak{g}^{\alpha}, \quad \underline{\mathfrak{u}} = \bigoplus_{\alpha \in \Delta(\mathfrak{h},\, \mathfrak{g}),\ \alpha(ih) > 0} \mathfrak{g}^{\alpha}, \quad (V.\ 4.7)$$

is possible to built to a connect subgroup an $\mathfrak{u}$-cohomology of Dolbeault to the complex context of the space of the generalized D-modules on $\mathfrak{q}$, whose class space G/Q, induce their structures on the open G_0-orbits D, then the generalized flag manifold given for G_0/L_0, where L_0, is the centralizator of a compact torus $\mathbb{T}$, (the which is reductive and connect like compact subgroup L), is the space of Borel subalgebras, which will include all the representations of

---

[2] For the cohomology theorem of the principal series exist a representation of finite dimension L, of M, such that

$$H^*(\mathfrak{g}, K; I \otimes L_s) = H^*(\mathfrak{m}, K \cap M; H \otimes L_s) \otimes \wedge \mathfrak{a}^*{}_C$$

with a variation grade of $l(s)$. Then the first factor of M, have concentrate cohomology in the interval $[q_0({}^0M), q_0({}^0M) + l_0({}^0M)]$. L_s, is a representation of a connect reductive group.

elliptic type[3] in G, (the Casimir operator act for 0 on the (L, $\mathfrak{l}$)-module of finite dimension $E_{s(\rho+\lambda)-\rho}$), and extensions of the fundamental series.

This can pre-write it intrinsically through of the sheaf of holomorphic functions $\mathfrak{o}$. Using the $\mathfrak{u}$-specialization of the Dolbeault cohomology[4] and considering L a connect component of the group G, corresponding to Lie algebra $\mathfrak{l}$, we can define the relative complex cohomology of Lie algebras given to the particular case L_0, like the space $\mathrm{Hom}((\wedge^{\bullet}\mathfrak{u}, C^{\infty}(G_0) \otimes V)^{L_0}$, always with the defined differentials for

$$df(X_1 \wedge \ldots \wedge x_n) = \sum_k (-1)^{k+1}\gamma(X_k)\{f(X_1 \wedge X_2 \wedge \ldots \wedge \bar{X}_k \wedge \ldots X_n)\} + \\ + \sum_{r<s} (-1)^{r+s} f([X_r, X_s] \wedge X_1 \wedge X_2 \wedge \ldots \wedge \bar{X}_r \wedge \ldots \wedge \bar{X}_s \wedge \ldots \wedge X_n) \qquad \text{(V. 4.8)}$$

where $X_k \in \mathfrak{u}$, k = 1, …, •, and $f \in \mathrm{Hom}((\wedge^{\bullet}\mathfrak{u}, C^{\infty}(G_0) \otimes V))^{L_0 \bullet}\gamma$, is the action of $\mathfrak{u}$, on $C^{\infty}(G_0) \otimes V$, through of the right translation $r \otimes \pi$. Extending this cohomology on all the component L then is necessary consider a space of infinite dimension representation, which must use the cohomological space of representations $H^j(G/L, \mathfrak{O}(E_{\chi}))$, to obtain the classification on the opened L- orbit on the flag manifold of Borel subalgebras of $\mathfrak{l} = \mathfrak{q}^r$, and the corresponding location through the Zuckerman functor $A^j(G, L, \mathfrak{q}; \chi)$ of the fundamental series (recovery the classic series or of Harish-Chandra) inside of the space of induced irreducible unitary representations. Then from it, we can to propose a theorem using the $\underline{\partial}$-cohomology of Schmid and the possibility of generalize the D-modules on an extension of continuous homomorphism between open orbits of a complex holomorphic bundle $G/L \rightarrow E_{\chi}$, and by this way to give an $\mathfrak{u}$-implicit classification with the corresponding Langlands data for the irreducible representations. Nevertheless the following problems arise: we can give a clear definition of the topology of the Dolbeault complex for this case and the problem of closed range of the operator $\underline{\partial}$-cohomology with the same infinitesimal character also of a vanishing theorem to the anti-dominating case. After we could give an analogous result to the case of infinite dimension considering representations on measurable orbits of flag manifolds.

Newly, considering the Casselman subrepresentations and let (σ, H), be a differentiable admissible Fréchet ${}^0M$-module with infinitesimal character χ_{σ}, and let $\nu \in \mathfrak{a}^*_c$. Then the induced representation $(\pi_{P,\sigma,\nu}, I_{P,\sigma,\nu})$, is the representation defined by the right translations on the defined ($\mathfrak{g}$, K)-module as the space

---

[3] All the canonical representations of L, whose character have support on an elliptic space in G. This subspace is a fundamental series representation class. The elliptic set in G, is the union of the G-conjugated of $T \cap G'$, where $H \times A$, is a fundamental Cartan subgroup, $T = H \cap G$, and G', a regular set.

[4] **Def. V. 1.** A $\mathfrak{u}$-specialization of the Dolbeault cohomology is the relative complex Lie algebras given for the space

$$A^{\bullet}(V) = \mathrm{Hom}_{L_0 \cap K_0}(\wedge^{\bullet}\mathfrak{u}, C^{\infty}(G_0) \otimes V)$$

where V is a complex $\mathfrak{g}$-module.

$$I_{P,\sigma,\nu} = \{f\in C^{\infty}(G, H_{\sigma}) | f(mang) = a^{\rho_P+\nu}\sigma(m)f(g),\ \forall\ a\in A,\ g\in G,\ n\in N \text{ and } m\in {}^0M\}, \quad (V. 4.9)$$

where using the Casselman notation

$$I_{P,\sigma,\nu} = Ind_P{}^G(H_{\sigma}\otimes C_{\rho P+\nu}), \quad (V. 4.10)$$

Then by the cohomology theorem of the principal series [], is clear that $\forall\lambda\in\mathfrak{h}^*c$, a dominate weight and F_{λ}, a simple Gc-module with parameter λ, some of the induced representations of some tempered representations are the given by the cohomological space of dimension $q\in\mathbb{N}$,

$$H^q(\mathfrak{g}, K; I\otimes F_{\lambda}) = 0, \quad (V. 4.11)$$

$\forall\ q\notin[q_0({}^0M) + l(s),\ q_0({}^0M) + l(s) + l_0({}^0M) + \dim A]$. Then using the defined isomorphism by the cohomology of the ${}^0M$-modules (like the obtained by the Casselman subrepresentations given in (5. 4. 4)) and their corresponding $\mathfrak{u}$-specialization we have the $\mathfrak{u}$-specialization given by the space

$$H^j(\mathfrak{u}, F) = \bigoplus_{s\in WP,\ l(s)=j} E_{s(\rho+\lambda)-\rho}, \quad (V. 4.12)$$

where F, is a irreducible $\mathfrak{g}$-module of finite dimension with maximum weight λ, and prescribed like $\mathfrak{l}$-module under this orthogonal composition. Then a possible generalization of the induced representations for (V. 4. 12), (in the sense of classify representations more fine inside of a global context (($\mathfrak{g}$, K)-*cohomology to the classification of a major class of fine representations*)) can be obtained using the Osborne lemma and the theorem of Casselman subrepresentation given by the identification

$$U(\mathfrak{u}^-)FU(\mathfrak{g})^H U(\mathfrak{u}) = U(\mathfrak{g}c), \quad (V. 4.13)$$

where if $j = n = dim\mathfrak{u}$, then in particular is had that

$$H^n(\mathfrak{u}, F) = \Lambda^n\mathfrak{u}^*\otimes F/\mathfrak{u}F, \quad (V. 4.14)$$

Now, the advantage of construct an extension of the space F, to the case of infinite dimension like in (V. 4. 13), reduce the evaluations followed topass of the Vogan-Zuckerman scheme to the D-modules scheme inside of the orbital context on all L. Also result clear through of this extension the followed way to prove the theorem of ($\mathfrak{g}$, K)-cohomology (Vogan-Zuckerman), through of the induction of the $\mathfrak{u}$-cohomology to all the reductive algebra $\mathfrak{g}$. Likewise, if $\mathfrak{q} = \mathfrak{b}$, and considering the equivalence between the categories of $U(\mathfrak{g})$-modules with infinitesimal character with the category of the D-modules on the flag manifolds of $\mathfrak{g}$, is possible deduce an analogous extension on the coherent sheaves space of the form

$$\mathcal{U}(\mathfrak{n})F_{\lambda}\mathcal{Z}(\mathfrak{g})^H\mathcal{U}(\mathfrak{h}) = \mathcal{U}(\mathfrak{g}), \quad (V. 4.15)$$

where F_λ, is the irreducible G-module of finite dimension belonging to an associative sheaf $\mathcal{D}_\lambda$, with maximum weight λ.

As establish Vogan in the study of the unipotent representations and cohomological induction of these representations, we can consider an irreducible unitary representation class of homogeneous space G/L, with Levi subgroup L, and determine through of a functorial process on $(\mathfrak{l}, L \cap K)$-modules of finite length, the induction to $(\mathfrak{g}, L \cap K)$-modules of the form

$$W_{\mathfrak{g}} \cong \mathrm{Ind}_{\mathfrak{g}, L \cap K} G/L, \qquad \text{(V. 4.16)}$$

and in all this process see the preservation of the Hermitian form of the Zuckerman functors on $W_{\mathfrak{g}}$. But this method is an orbital version of the construct of representations through of parabolic induction (which use $\mathfrak{n}$-cohomology to the obtaining of unitarizable irreducible modules) considering hypothesis on the polarization $\mathfrak{q}$. Nevertheless, this restriction not establish a criteria of classification to the very infinitesimal numbering characters that appears and their relation with the co-adjunct nilpotent orbits, thus is necessary the application of the minimal K-types of Salamanca-Vogan. This via can be the algebraic way to the solution of the problem of classification of all the irreducible unitary representations. Nevertheless only can get us partial solutions, stayed several aisled representations in this big class of representations.

Is clear that when G, is non-compact and L, is compact the representations are the identified by the Langlands conjecture, which was proved by Schmid. To L, non-compact the analytic situation is no clear and to it, the via that is used is consider only the cocycles $C^{0,s}(K/(K \cap L))$ (*Space of the* (0, s)-*forms where* $K/K \cap L$, *are the* s-*dimensional complex compact submanifolds of* G/L), of the space $H^{0,s}(K/(K \cap L), \mathbb{C}_{\lambda+2\delta(\mathfrak{u})})$, that is an $\mathfrak{u}$-specialization of the cohomological space of the Borel-Weil theorem, $H^{0,q}(G/L, V)$, and we determine the Penrose transform on said cocycles. But the identification of irreducible unitary representations with λ, appropriate result very difficult, since as is told above of this exposition is not result clear that the co-boundary operator $\underline{\partial}$, on the images of the Dolbeault complexes under the Penrose transform belonging to space $\mathcal{P}(H^{0,s}(K/(K \cap L), \mathbb{C}_{\lambda+2\delta(\mathfrak{u})}))$ be of closed range.

Chapter VI

Intertwinning Operators

VI. 1.

In the following exposition will establishes some consequent properties of the behavior of the corresponding principal series to the temperate representation theory. Thus of it, will studies some integral identities such that

$$\mathrm{Lim}_{\tau\to\infty} a^{\mu+\rho(h)} \langle \pi(\exp(th)f, g)\rangle = \int_{\underline{N}F} \langle f(\underline{n}), g(1)\rangle d\underline{n}, \ \forall \underline{n} \in \underline{N}_F,$$

that will use to establish an important criteria to the Langlands classification of the tempered (g, K)-modules belonging to the class $L^2(G)$, on a real reductive group G.

Consider $F \subset \Delta_0$, and let (P_F, A_F), be corresponding canonical parabolic pair. We fix to (σ, H_σ), like a representation of ${}^0M_F$, such that satisfies the weak inequality. Let $N_F = \theta(N_F)$, $K_F = K \cap M_F$, and we define explicitly to the root system

$$\Phi(P_F, A_F) = \{\alpha \in \Phi(P, A) \mid \alpha|_{\mathfrak{a}_F} = 0\}, \qquad \text{(VI. 1)}$$

that is equivalent to say that the root system $\Phi(P_F, A_F)$, denotes the root set of $\mathfrak{a}_F$, on $\mathfrak{n}_F$.

Lemma VI. 1. Let $\mu \in (\mathfrak{a}_F)_C^*$, such that $\mathrm{Re}(\mu, \alpha) > 0$, to any $\alpha \in \Phi(P_F, A_F)$.

1. If $f \in (H^{\sigma, \mu})_\infty$, and if $w \in (H_\sigma)_K$, then

$$\int_{\underline{N}F} |\langle f(\underline{n}), w\rangle| d\underline{n} < \infty, \qquad \text{(VI. 2)}$$

with major reason the map

$$f \mapsto \int_{\underline{N}F} |\langle f(\underline{n}), w\rangle| d\underline{n}, \qquad \text{(VI. 3)}$$

is continuous on $(H^{\sigma, \mu})_\infty$.

2. If $w \in (H_\sigma)_K$, is not null then exist $f \in I_{PF, \sigma, \mu}$, such that $\int_{\underline{N}F} |\langle f(\underline{n}), w\rangle| d\underline{n}$, is not null.

Proof. Consider $\underline{n} \in \underline{N}_F$, then $f(\underline{n}) = f(n m_F(\underline{n}) a_F(\underline{n}) k_F(\underline{n}))$. Then

$$f(n m_F(\underline{n}) a_F(\underline{n}) k_F(\underline{n})) = a_F(\underline{n})^{\mu+\rho F} \sigma(m_F(\underline{n}) f(k_F(\underline{n})) = f(\underline{n}),$$

Then

INTECH
open science | open minds

$$\int_{\underline{N}F}|\langle f(\underline{n}), w\rangle|d\underline{n} = \int_{\underline{N}F} a_F(\underline{n})^{Re\mu+\rho}|\sigma(m_F(\underline{n})f(k_F(\underline{n})), w\rangle|d\underline{n},$$

But

$$\langle\sigma(m_F(\underline{n})f(k_F(\underline{n})), w\rangle \leq \beta(f(k_F(\underline{n}))(1 + \log||m_F(\underline{n})||)^r\Xi_F(m_F(\underline{n})),$$

then

$$\int_{\underline{N}F}|\langle f(\underline{n}), w\rangle|d\underline{n} \leq \int_{\underline{N}F} \beta(f(k_F(\underline{n}))a_F(\underline{n})^{\mu+\rho}(1 + \log||m_F(\underline{n})||)^r\Xi_F(m_F(\underline{n}))d\underline{n},$$

with β, a continuous semi-norm on the space $(H_\sigma)^\infty$, where has used the weak inequality. We define $\gamma(f) = \sup_{k\in K}\beta(f(k))$. Now, by a lemma of asymptotic boundess (consult [Bulnes, F, 2002]), we have

$$a_F(\underline{n})^{Re\mu+\rho} \leq C_q(1 - \rho(\log a_F(\underline{n}))^{-q}, \forall\ q > 0, \tag{VI. 4}$$

Then

$$\beta(f(k_F(\underline{n})a_F(\underline{n})a_F(\underline{n})^{\mu+\rho}(1 + \log||m_F(\underline{n})||)^r\Xi_F(m_F(\underline{n})) <$$

$$< C_q\gamma(f)a_F(\underline{n})\Xi_F(m_F(\underline{n}))(1 - \rho\log a_F(\underline{n}))^{-q},$$

$\forall\ q > 0$. Then given that

$$\int_{\underline{N}F} C_q\gamma(f)a_F(\underline{n})\Xi_F(m_F(\underline{n}))(1 - \rho\log a_F(\underline{n}))^{-q}d\underline{n} < \infty,$$

Then with major reason $\int_{\underline{N}F}|\langle f(\underline{n}), w\rangle|d\underline{n} < \infty$. Then stay demonstrated (1). To demonstrate that the map defined by the correspondences

$$f \mapsto \int_{\underline{N}F}|\langle f(\underline{n}), w\rangle|d\underline{n},$$

is continuous on the space $(H^{\sigma,\mu})_\infty$, is only necessary see that $\int_{\underline{N}F}|\langle f(\underline{n}), w\rangle|d\underline{n} \leq \int_{\underline{N}F} \beta(f(k_F(\underline{n}))a_F(\underline{n})^{\mu+\rho}(1 + \log||m_F(\underline{n})||)^r\Xi_F(m_F(\underline{n}))d\underline{n}$, where

$$f(\underline{n}) = a_F(\underline{n})^{\mu+\rho_F}\sigma(m_F(\underline{n}))f(k_F(\underline{n})),$$

that is to say, the topology of $f(\underline{n})$, is the induced for the semi-norms $\sigma(m_F(\underline{n}))$, on ${}^0M_F$, in the decomposition of Langlands ${}^0MAN$.

In particular, all p-canonical pair has an infinitesimal equivalent to $(\sigma, (H^{\sigma,\mu})^\infty)$ induced to a representation on $(H^{\sigma,\mu})_\infty$. Then the induced topology is the semi-norms topology in $(H^{\sigma,\mu})_\infty$. Thus the map given by (VI. 3), is continuous.

We consider now a partition of the unity in $\underline{N}_F$, that is to say, a family of functions $\{H_i\}_{i\in I} \subset C_c(\underline{N}_F)$, such that

$$H(\underline{n}) = \begin{cases} 1, & if \quad \underline{n} \in \underline{N_F} \\ 0, & if \quad \underline{n} \notin N_F \end{cases} \tag{VI. 5}$$

$\forall\, H(\underline{n}) \in C_c(\underline{N}_F)$. Since

$$f(nma\underline{n}) = a^{\mu+\rho}\sigma(m)H(\underline{n})w = <\sigma(m)H(\underline{n}), w>, \tag{VI. 6}$$

$\forall \underline{n} \in \underline{N}_F$, $m \in {}^0M_F$, $a \in A_F$, and $n \in N_F$, and due to that the topology with such partition of the unity is equivalent to the induced for the semi-norms in $(H^{\sigma,\mu})_\infty$, since $f(nma\underline{n}) = <\sigma(m)H(\underline{n}), w>$, is continuous in the same sense that

$$f(nm_F(\underline{n})a_F(\underline{n})k_F(\underline{n})) = a_F(\underline{n})^{\mu+\rho_F}\sigma(m_F(\underline{n}))f(k_F(\underline{n})),$$

in $H^{PF,\sigma,\mu}$ (that is to say, $f(\underline{n}) = f(nm_F(\underline{n})a_F(\underline{n})k_F(\underline{n}))$, $\forall \underline{n} \in \underline{N}_F$), then $f \in (H^{\sigma,\mu})_\infty$, and

$$\int_{\underline{N}F} |<f(\underline{n}), w>| d\underline{n} = <w, w>> 0,$$

Being $I_{PF,\sigma,\mu}$, dense in $(H^{\sigma,\mu})_\infty$ then $\forall\, f \in I_{PF,\sigma,\mu}$, $\int_{\underline{N}F} |<f(\underline{n}), w>| d\underline{n}$, is not null. ■

Lemma VI. 2. Let $f \in I_{PF,\sigma,\mu}$, be then there is a subspace of finite dimension V(f), of $(H_\sigma)_K$, such that

$$\int_{\underline{N}F} <f(\underline{n}), w> d\underline{n} = 0,$$

$\forall\, w \in (H_\sigma)_K$, orthogonal to V(f).

Proof. Let $w \in (H_\sigma)_K$. If $k \in K_F$, then

$$\int_{\underline{N}F} <f(\underline{n}), w> d\underline{n} = \int_{\underline{N}F} <\sigma(k)f(k^{-1}\underline{n}k), w> d\underline{n} = \int_{\underline{N}F} <\sigma(k)f(\underline{n}), w> d\underline{n}$$

$$= \int_{\underline{N}F} <f(\underline{n}), \sigma(k^{-1})w> d\underline{n},$$

where has been used the invariance of $d\underline{n}$, on $\underline{N}_F$, under conjugation for k_F. Let

$$S = \{\gamma\}_{\gamma \in K^\wedge}, \tag{VI. 7}$$

and we consider

$$V(f) = \sum_{\delta \in S,\, \delta \in [\gamma] \subset K^\wedge} H_\sigma(\delta), \tag{VI. 8}$$

since S, is finite $\forall \gamma \in S$, then

$$\dim H_\sigma(\delta)f < \infty, \ \forall \ f \in V(f), \tag{VI. 9}$$

then dim $V(f) < \infty$. Then since

$$\int_{\underline{N}F} \langle f(\underline{n}), w \rangle d\underline{n} = \langle V(f), w \rangle,$$

$\forall \ f \in (H^{\sigma, \mu})^\infty$. Then

$$\int_{\underline{N}F} \langle f(\underline{n}), w \rangle d\underline{n} = 0$$

If and only if $(H_\sigma)^0{}_K \perp V(f)$, with $(H_\sigma)^0{}_K = \{w\} \subset (H_\sigma)_K$. ■

Now well, since

$$\int_{\underline{N}F} \langle f(\underline{n}), w \rangle d\underline{n} = \langle V(f), w \rangle,$$

Then exist a map

$$\beta_{PF, \sigma, \mu}: I_{PF, \sigma, \mu} \rightarrow (H_\sigma)_K, \tag{VI. 10}$$

whose correspondence rule is

$$f \mapsto \beta_{PF, \sigma, \mu}(f)w = \int_{\underline{N}F} \langle f(\underline{n}), w \rangle d\underline{n}, \tag{VI. 11}$$

Since $\forall \ k \in K_F$, and $\forall \underline{n} \in \underline{N}_F$,

$$\beta_{PF, \sigma, \mu}(f(\underline{n}k))w = \beta_{PF, \sigma, \mu}(\sigma(k)f(k^{-1}\underline{n}k))w$$

$$= \beta_{PF, \sigma, \mu}(\sigma(k)f(\underline{n}))w$$

$$= \beta_{PF, \sigma, \mu}(f(\underline{n})\sigma(k^{-1}))w, \ \forall \ w \in (H_\sigma)_K,$$

Then $\beta_{PF, \sigma, \mu}$, is an invariant homomorphism under the conjugation of K_F. Being the $({}^0M_F, K)$-modules $I_{PF, \sigma, \mu}$, and $(H_\sigma)_K$, K_F- modules, the homomorphism $\beta_{PF, \sigma, \mu}$, is a homomorphism of K_F-modules.

Lemma VI. 3. Let $\beta_{PF, \sigma, \mu}(\underline{\mathfrak{n}}_F, I_{PF, \sigma, \mu}) = 0$. Let $\alpha_{PF, \sigma, \mu}$, be the corresponding linear map

$$\alpha_{PF, \sigma, \mu}: I_{PF, \sigma, \mu}/\underline{\mathfrak{n}}_F I_{PF, \sigma, \mu} \rightarrow (H_\sigma)_K, \tag{VI. 12}$$

whose rule of correspondence is

$$f \mapsto \alpha_{PF, \sigma, \mu}(f)w = \int_{\underline{N}F} \langle f(\underline{n}), w \rangle d\underline{n},$$

then $\alpha_{PF, \sigma, \mu} \in Hom_{\mathfrak{m}F, K}(I_{PF, \sigma, \mu}/\underline{\mathfrak{n}}_F I_{PF, \sigma, \mu}, (H_\sigma)_K)$.

Proof. Indeed, consider $f \in I_{PF,\sigma,\mu}$, and let $X \in \mathfrak{n}_F$. If $w \in (H_\sigma)_K$, then we can designate

$$\gamma_w(f) = \int_{\underline{N}F} \langle f(\underline{n}), w \rangle d\underline{n},$$

Since γ_w, is a defined functional in $I_{PF,\sigma,\mu}$, and this is a dense space in $(H^{\sigma,\mu})_\infty$, are the induced by the same numerable space of semi-norms, then γ_w, is a continuous functional. Then $\forall$ $X \in \mathfrak{n}_F$,

$$\gamma_w(Xf) = d/dt\, \gamma_w(\pi_{\sigma,\mu}(\exp(tX)f)|_{t=0} = 0, \qquad \text{(VI. 13)}$$

Which is true for the right invariance of $d\underline{n}$, on $\underline{N}_F$, (then $\beta_{PF,\sigma,\mu}(\underline{\mathfrak{n}}_F, I_{PF,\sigma,\mu}) = 0$). If $X \in {}^0\mathfrak{m}_F$, then for the invariance under the conjugation of elements ${}^0M_F$, is had that:

$$\gamma_w(Xf) = d/dt\, \gamma_w(\pi_{\sigma,\mu}(\exp(tX)f)|_{t=0} = (d/dt)\int_{\underline{N}F} \langle f(\underline{n}\exp(tX)), w \rangle d\underline{n}|_{t=0}$$

$$= (d/dt)\int_{\underline{N}F} \langle \sigma(\exp(tX))f(\exp(-tX))\underline{n}\exp(tX)), w \rangle d\underline{n}|_{t=0}$$

$$= (d/dt)\int_{\underline{N}F} \langle \sigma(\exp(tX))f(\underline{n}), w \rangle d\underline{n},$$

where the last equation is followed of the invariance of the measure $d\underline{n}$, on $\underline{N}_F$, under conjugation by elements of ${}^0M_F$. Since there is a subspace of finite dimension such that $\forall$ $w \in (H_\sigma)_K$, $\{w\}$, is orthogonal to said subspace of $I_{PF,\sigma,\mu}$, then $\beta_{PF,\sigma,\mu}(\underline{\mathfrak{n}}_F, I_{PF,\sigma,\mu}) = 0$, and is ${}^0\mathfrak{m}_F$-invariant, that is to say,

$$\beta_{PF,\sigma,\mu}(Xf) = X\beta_{PF,\sigma,\mu}(f), \quad \forall \quad X \in {}^0\mathfrak{m}_F, \qquad \text{(VI. 14)}$$

If $h \in \mathfrak{a}_F$, then is necessary consider in the K_F- homomorphism $\beta_{PF,\sigma,\mu}$, the following $\underline{N}_F$-measure

$$d(a\underline{n}a^{-1}) = a^{-2\rho_F} d\underline{n}, \qquad \text{(VI. 15)}$$

and find that under conjugation of K_F, satisfies that

$$\beta_{PF,\sigma,\mu}(hf) = (\mu - \rho_F)(h)\beta_{PF,\sigma,\mu}(f) \in (H_{\sigma,\mu-\rho^F})_K,$$

Thus $\alpha_{PF,\sigma,\mu} \in Hom_{\mathfrak{m}F,K}(I_{PF,\sigma,\mu}/\underline{\mathfrak{n}}_F I_{PF,\sigma,\mu}, (H_\sigma)_K)$. ■

The lemma VI. 3., jointly with the lemma on the fast decreasing of a temperate irreducible representation of G, of integrable square (Conference Vol. 1 temperate ($\mathfrak{g}$, K)-modules [28]) implies that exist a homomorphism of ($\mathfrak{g}$, K)-modules

$$j_{PF,\sigma,\mu}: I_{PF,\sigma,\mu} \rightarrow I_{PF,\sigma,\mu}, \qquad \text{VI. 16)}$$

whose rule of correspondence is

$$f \mapsto \alpha_{PF,\sigma,\mu}(f) = j_{PF,\sigma,\mu}(f)(1), \qquad \text{(VI. 17)}$$

and since $I_{P_F, \sigma, \mu}$, is a ($\mathfrak{g}$, ${}^0M$)-module non-vanishing then $j_{P_F, \sigma, \mu}$, is non-vanishing by the lemma on fast decreasing of a temperate irreducible representation belonging to the space $L^2(G)$. This last will be an important question in the after applications of the representation theory.

Theorem VI. 1. Maintaining the previous affirmations, let $f \in (H^{\sigma, \mu})_{\infty}$, and $g \in I_{P_F, \sigma, \mu}$. Let $h \in \mathfrak{a}_F$, be such that $\alpha(h) > 0$, $\forall \alpha \in \Phi(P_F, A_F)$, then

$$\lim_{t \to +\infty} e^{t(\rho - \mu)(h)} \langle \pi(\exp(th))f, g \rangle = \int_{\underline{N}_F} \langle f(\underline{n}), g(1) \rangle d\underline{n}, \qquad \text{(VI. 18)}$$

where $\pi = \pi_{P_F, \sigma, \mu}$.

Proof. For one side g, is K-finite and by the lemma on the finitely generated [5, 28, 40], the generated by g(K), is of finite dimension. Since $f \in (H^{\sigma, \mu})_{\infty}$, then $f(g) \in (H_\sigma)^{\infty}$, $\forall\ g \in G$. Consider $a_t = \exp(th)$, $\forall\ h \in \mathfrak{a}_F$, then

$$\langle \pi(\exp(th))f, g \rangle = \langle \pi(\exp(a_t))f, g \rangle = \int_K \langle f(ka_t), g(k) \rangle dk, \quad \forall\ k \in K \qquad \text{(VI. 19)}$$

But, since the measure on G (K-invariant) can be normalized then by the theory of integration on parabolic pairs [Wallach] is had that

$$\int_K \langle f(ka_t), g(k) \rangle dk = \int_{\underline{N}_F} a_F(\underline{n})^{2\rho} \langle f(k(\underline{n})a_t), g(k(\underline{n})) \rangle d\underline{n},$$

$\forall\ k(\underline{n})a_t \in K_F$, $k(\underline{n}) \in K$, and $\underline{n} \in N_F$. Now well, $n = n m_F(\underline{n}) a_F(\underline{n}) k_F(\underline{n})$ $\forall \underline{n} \in N_F$. Consider $k_F(\underline{n}) \in N_F(m_F(\underline{n}) a_F(\underline{n}))^{-1}\underline{n}$. This circumstance will imply to the measure dg that:

$$\langle \pi(\exp(a_t))f, g \rangle = \int_{\underline{N}_F} a_F(\underline{n})^{2\rho} a_F(\underline{n})^{-\rho-\mu} \langle \sigma(m_F(\underline{n}))^{-1} f(\underline{n}a_t), g(k(\underline{n})) \rangle d\underline{n}$$

$$= \int_{\underline{N}_F} a_F(a_t\underline{n}a_{-t})^{2\rho} a_F(a_t\underline{n}a_{-t})^{-\rho-\mu} \langle \sigma(m_F(\underline{n}))^{-1} f(a_t\underline{n}a_{-t}), g(k(\underline{n})) \rangle d\underline{n}$$

$$= \int_{\underline{N}_F} a_t^{\rho+\mu} a_F(\underline{n})^{\rho-\mu} \langle \sigma(m_F(\underline{n}))^{-1} f(a_t\underline{n}a_{-t}), g(k(\underline{n})) \rangle d\underline{n}$$

$$= a_t^{\rho+\mu} \int_{\underline{N}_F} a_F(\underline{n})^{\rho-\mu} \langle \sigma(m_F(\underline{n}))^{-1} f(a_t\underline{n}a_{-t}), g(k(\underline{n})) \rangle d\underline{n},$$

where has been considered that

$$a_t^{\rho+\mu} \Xi(a_t) = \int_{\underline{N}_F} a_F(\underline{n})^{\rho-\mu} a(a_t\underline{n}a_{-t})^{\rho} d\underline{n}, \qquad \text{(VI. 20)}$$

where $a(a_t\underline{n}a_{-t})^{\rho} = \sigma(m_F(\underline{n}))^{-1} f(a_t\underline{n}a_{-t})$, $\forall \underline{n} \in N_F$. Then

$$\langle \pi(\exp(th))f, g \rangle = a_t^{\rho+\mu} \int_{\underline{N}_F} a_F(\underline{n})^{\rho-\mu} \langle \sigma(m_F(\underline{n}))^{-1} f(\underline{n}), g(k(a_t\underline{n}a_{-t})) \rangle d\underline{n}, \qquad \text{(VI. 21)}$$

If is possible to exchange the limit for the integral, the result will follows trivially. For this goal, we consider a measurable subset E, of N, to know;

$$I_t(E) = \int_E a_F(a_t\underline{n}a_{-t})^{\rho-\mu}<\sigma(m_F(a_t\underline{n}a_{-t}))^{-1}f(\underline{n}), g(k(a_t\underline{n}a_{-t}))>d\underline{n}, \qquad \text{(VI. 22)}$$

that is to say, we will elect a measurable subset E, of N, whose integral is defined under said rule of correspondence and under the normalization of their measure.

By this way, one can to demonstrate that exist an integrable function u, on N_F, such that

$$I_t(E) \leq \int_E u(\underline{n})d\underline{n}, \ \forall \ t > 0, \qquad \text{(VI. 23)}$$

In this way, the justification of the interchange of the limit for the integral is a mere consequence of the Vitali theorem. Likewise, the rule of transformation of f implies that

$$I_t(E) = \int_E a_F(a_t\underline{n}a_{-t})^{\rho-\mu}a_F(\underline{n})^{\rho+\mu}<\sigma(m_F(a_t\underline{n}a_{-t}))^{-1}m_F(\underline{n})f(\underline{n}), g(k(a_t\underline{n}a_{-t}))>d\underline{n}$$

where

$$a_F(a_t\underline{n}a_{-t})^{\rho-\mu}a_F(\underline{n})^{\rho+\mu}\,|<\sigma(m_F(a_t\underline{n}a_{-t}))^{-1}m_F(n)f(\underline{n}), g(k(a_t\underline{n}a_{-t}))>| \leq$$

$$\leq a_F(a_t\underline{n}a_{-t})^{\rho-\mathrm{Re}\mu}\Xi_F(m_F(a_t\underline{n}a_{-t}))^{-1}m_F(\underline{n})a_F(\underline{n})^{\rho+\mathrm{Re}\mu}(1 + \log\| m_F(\underline{n}) \|)^d,$$

Thus the expression of the limit followed of the integral $I_t(E)$, result be finite and bounded by the integral (VI. 23). Then analyzing the expression of the function Ξ_F, evaluated in$m_F(a_t\underline{n}a_{-t}))^{-1}m_F(\underline{n})\in{}^0M_F$, is applied the property to this case, following:

$$\Xi_F (x^{-1}y) = \int_{K_F} a(kx)^{\rho}a(ky)^{\rho}dk, \ \forall \ x, y\in{}^0M_F, \ \text{(I)}$$

signing

$$\Xi_F (x^{-1}y) = \int_{K_F} a(kx^{-1}y)dk = \int_{K_F} a(k(kx)x^{-1}y)a(kx)^{2\rho}dk$$

$$= \int_{K_F} a(ky)^{\rho}a(kx)^{\rho}dk, \quad \forall \ x, y\in{}^0M_F,$$

where kx = n(kx)a(kx)k(kx). Then

$$\int_{K_F} a(ky)^{\rho}a(kx)^{\rho}dk = \int_{{}^*\underline{N}_F} a(k(\underline{n})x)^{\rho}a(k(\underline{n})y)^{\rho}a(n)^{2\rho}d\underline{n},$$

The previous orbital integral imply to the inequality $I_t(E) \leq \int_E u(\underline{n})d\underline{n}, \ \forall \ t > 0$, that

$$I_t(E) \leq \int_{E\times{}^*\underline{N}_F} a_F(a_t\underline{n}a_{-t})^{\rho-\mathrm{Re}\mu}a({}^*\underline{n}m_F(\underline{n}))^{\rho}v(\underline{n})^{\rho}a_F(\underline{n})^{\rho+\mathrm{Re}\mu}d\underline{n}d\underline{n}^*,$$

with $v(\underline{n}) = (1 + \log\| m_F(\underline{n}) \|)^d$. Now, considering $a(m_F(g))a_F(g) = a(g)$, and

$${}^*\underline{n}m_F(a_t\underline{n}a_{-t}) = m_F(a^*{}_t\underline{n}a_{-t}),$$

then

$$I_t(E) \leq \int_{E \times {}^*\underline{N}_F} a({}^*\underline{n}a_t\underline{n}a_{-t})^{\rho - \mathrm{Re}\mu} a({}^*\underline{nn})^{\rho + \mathrm{Re}\mu} v(\underline{n}) d\underline{n} d\underline{n}^*,$$

But due to the K-invariance of the measure dg, normalized in N_F, then

$$\int_{E \times {}^*\underline{N}_F} a({}^*\underline{n}a_t\underline{n}a_{-t})^{\rho - \mathrm{Re}\mu} a({}^*\underline{nn})^{\rho + \mathrm{Re}\mu} v(\underline{n}) d\underline{n} d\underline{n}^*$$

$$= \int_{E \times {}^*\underline{N}_F} a(k({}^*\underline{n})a_t\underline{n}a_{-t})^{\rho - \mathrm{Re}\mu} a(k({}^*\underline{n})\underline{n})^{\rho + \mathrm{Re}\mu} v(\underline{n}) a({}^*\underline{n})^{2\rho} d\underline{n} d\underline{n}^*$$

$$= \int_{E \times {}^*\underline{N}_F} a(a_t k({}^*\underline{n})\underline{n}k({}^*\underline{n})a_{-t})^{\rho - \mathrm{Re}\mu} a(k({}^*\underline{n})\underline{n}k({}^*\underline{n}))^{-1})^{\rho + \mathrm{Re}\mu} v(\underline{n}) a({}^*\underline{n})^{2\rho} d\underline{n}^* d\underline{n}$$

$$= \int_{E \times {}^*\underline{N}_F} a({}^*\underline{n})^{2\rho} a(a_t\underline{n}a_{-t})^{\rho - \mathrm{Re}\mu} a(\underline{n})^{\rho + \mathrm{Re}\mu} (1 + \log \| m_F(\underline{n}) \|)^d d\underline{n}^* d\underline{n},$$

$$= \int_E a(a_t\underline{n}a_{-t})^{\rho - \mathrm{Re}\mu} a(\underline{n})^{\rho + \mathrm{Re}\mu} (1 + \log \| m_F(\underline{n}) \|)^d d\underline{n},$$

Let $0 < \varepsilon < 1$, be, such that $\langle \mathrm{Re}\mu - \varepsilon\rho_F, \alpha \rangle > 0$, $\forall \alpha \in \Phi(P_F, A_F)$. Then the kernel $a(a_t\underline{n}a_{-t})^{\rho - \mathrm{Re}\mu} a(\underline{n})^{\rho + \mathrm{Re}\mu}$, takes the form

$$a(a_t\underline{n}a_{-t})^{\rho - \mathrm{Re}\mu} a(\underline{n})^{\rho + \mathrm{Re}\mu} = a(a_t\underline{n}a_{-t})^{\rho - \varepsilon\rho_F} a(a_t\underline{n}a_{-t})^{-(\mathrm{Re}\mu - \varepsilon\rho_F)} a(\underline{n})^{\mathrm{Re}\mu - \varepsilon\rho_F} a(\underline{n})^{\rho + \varepsilon\rho_F} \leq a(\underline{n})^{\rho + \varepsilon\rho_F}, \text{ [5, 28]},$$

Considering the appendix C, is natural suppose that to any $q > 0$, there is $C_q > 0$ such that

$$a(\underline{n})^{\rho + \varepsilon\rho_F} \leq C_q a(\underline{n})^{\rho} (1 - \log \| a(\underline{n}) \|)^{-q}, \tag{VI. 24}$$

Considering to $u(\underline{n}) = a(\underline{n})^{\rho + \varepsilon\rho_F}(1 + \log \| \underline{n} \|)^d$, is had that u, is integrable on $\underline{N}$, and then $\forall$ E measurable subset of N, one satisfies that (VI. 23). Therefore the integral (VI. 19) exist, and since the limit exist then the Haar measure on N_F, is reduced to the integral

$$I = \int_{\underline{N}_F} \langle f(n), g(1) \rangle d\underline{n}, \ \forall \underline{n} \in \underline{N}_F. \ \blacksquare$$

Some Examples of Orbital Integrals into Representation Theory on Field Theory and Integral Geometry

VII. 1.

The development of orbital integrals into of the mathematical physics, conclave to use of integral representations that are realizations of certain unitary representations of Lie groups such as U(n), SU(2), SU(n), SU(2, 2), SU(p, q),and U(p, q), and that helps to obtain general solutions of partial differential equations in context of an algebra, like for example a quaternion algebra. In the case of establish equivalences through of a moduli space that can be descript by algebraic relations among geometrical identities, as the of quaternion analysis, carry us to establish a structural equivalence among the spaces $\mathbb{R}^4 \cong \mathbb{C}^2$, that which conclave to the re-definition in a context of hyper-complex analysis of vector objects in $\mathbb{R}^4$. Other example that we find in the integrals of the integral geometry is the given to $G_{2,4}(\mathbb{C})$, through of the space $\mathbb{P}^3(\mathbb{C})$, which establish a twistor geometry.

But the relations of isomorphism in vector homogeneous bundles are similar to the follows in integral developmental operator cohomology in vector tomography. Now, the intertwining integral operators among cohomological classes of both context (respective cycles), result be equivalent in context of bundles of lines (with equivalent cocycles). Then some conjectures that can be possible of prove to the light of some integral operators that establish this isomorphism is:

Conjecture VII. 1."The Penrose transform is a vector Radon transform on sections of homogeneous vector bundles in **M**′, being **M**, a complex Riemannian manifold.

And considering a reconstruct of a Riemannian manifold through of cycles, we can consider the following conjecture.

Conjecture VII. 2. "The twistor transform is the generalization of the Radon transform n Bundles of lines and the Penrose transform is a specialization of the twistor transform in S^4"

Of fact, the twistor transform can determines by the pair of Penrose transform on the G-orbits $\mathbb{P}^+$, y $\mathbb{P}^-$, [32, 33]. This is analogue to determine the Fourier transform in $\mathbb{R}^n$, like the combination of the Radon transform with the unidimensional Fourier transform calculated these also on Euclidean G-orbits[32, 33, 37], (hyperplanes and hyperlines).

In a more general sense, the vector bundles to the that it is does allusion are the seated in homogeneous spaces G^C/L, with homogeneous subspaces G^C/Q, where persist a complex holomorphic manifold with Q-orbits (parabolic orbits), of G^C/L, which have a Haar induced measure in every close submanifold given for the flags of the corresponding holomorphic vector G-bundles, and whose orbital integrals are evaluated on said orbits obtaining applicable invariants to all the homogeneous space in question, G^C/L. This process called of orbitalization of the homogeneous space G^C/L, is the base in major part in the inclusion of homogeneous spaces obtained for reduction of G^C/L, through their holonomy.

Likewise, calculating the orbital integrals on the corresponding orbits to the reductive homogeneous spaces obtained by reduction of G^C/L, that is to say; on the subjacent orbits in each one of the components of the sequence of inclusions [39]:

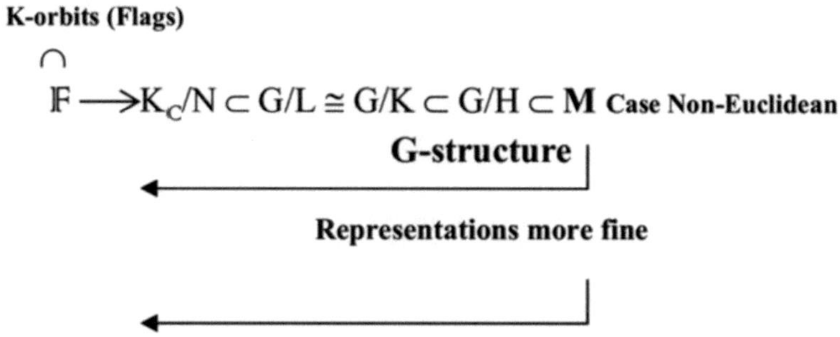

(VII. 1)

We obtain a constant operator in the applications of a homogeneous space in other, given by the complex integral operator on cohomological classes $H^s(D, O(\Lambda))$, where D, is the set of certain positive lines in $\mathbb{P}^n$, and whose image is the group of cohomology $H^s(G/H, \nu_n)$. ν_n represent the corresponding homogeneous bundle on the orbit $\mathbf{M}^+$, and s, is the complex dimension of **M**. The lines in $\mathbb{P}^n$, is identified in every case like the corresponding trajectories in **M**, to know,

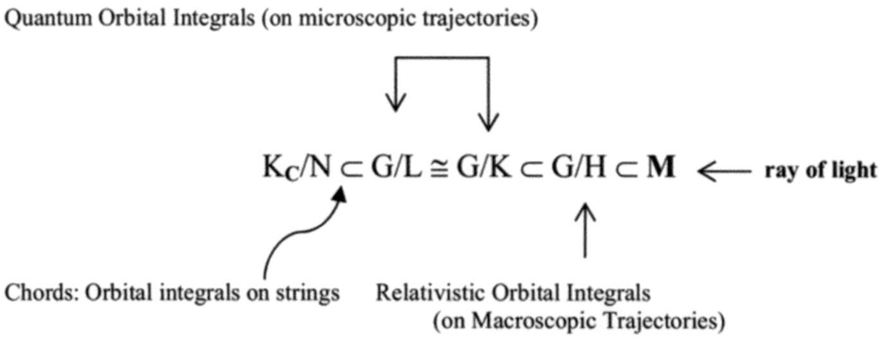

(VII. 2)

Of this way, a singularity in the G-structure of a differentiable manifold **M**, is a zero in the D_h-structure of **M**. This is precisely one of the developmental aspects by Penrose in their twistor theory through of the integral geometry to the manipulation of the infinites. D_h, is the space of lines in $\mathbb{P}^n$, whose image under certain integral transform is the space of points $\mathbb{P}^{n*}$, (the dual projective space to $\mathbb{P}^n$), and whose dual natural pairing of homogeneous coordinates is vanished in the context of the complex bundle of lines on $\mathcal{L}$.

The spaces $H^s(G/H, v_n)$, result be *fine representations*, of G/L, through of the bundle of flags corresponding to complex holomorphic G-bundle seated in G/L. The realizations of these unitary representations are the orbital integrals on the flags that are the K-orbits of the corresponding vector G-bundle. If we consider the corresponding Lie algebras to whose homogeneous spaces, in the frame of homogeneous space G/P, the representation spaces can be obtained and classified in the via of the *L^2-cohomología*, on parabolic orbits and through the corresponding orbital integrals.

This is precisely used in unitary representations, that is to say; the use of representations of G, (a complex Lie group), constructed on spaces of holomorphic sections of vector bundles and generalizations. The problem that here arises is the validation and verify of the *Hermitian forms* calculated through of the diverse integral realizations with the corresponding representations electing the *differentiable and adequate minimal globalization.* To it, we must use a general version of complex cohomology called *topological cohomology of Dolbeault*that is a *Serre generalization,* of the Dolbeault cohomology. The intertwining integrals in this case include *Szëgo operators* and others operators on *quotients of Langlands.*

A particular applied case to the context of the compact groups, and thus of the L^2-cohomology more simple (a subquotient of the L^2-cohomology on G/L), is that the holomorphic vector G-bundle of G/L $\cong$**M**, is isomorphic to vector bundle of G/K, with $L = Z_G(\mathbb{T})$, with $\mathbb{T}$, a maximal torus of G. The integral operators in this case results be Fourier transforms in the context of the complex holomorphic bundles. The lines in this case, result be the circles S^1, in $\mathbb{C}^n$.

The obtaining of an analytic function of a space to other, is realized of agree to the integral cohomology, $(n-1-q)$-$\bar{\partial}$-cohomology $H^{n-1,\,q}(\mathbb{C}^n/D, V)$, where D, is a linearly concave domain in $\mathbb{C}^n$. Likewise for example, if we consider the space $\mathbb{C}^2$, the integral operator that obtain a harmonic function in $\mathbb{R}^4$, through of to calculate functions of $C^\infty(\mathbb{C}^3)$,on lines S^1,in $\mathbb{C}^2$, comes given for

$$\phi(w, x, y, z) = \int_{S1} f((w+x)+(y+iz)\zeta, (y-iz)+(w-x)\zeta, \zeta)\, d\zeta, \qquad \text{(VII. 3)}$$

which have their geometrical re-interpretation like a Radon transform on lines of a flag manifold $\mathbb{F} = \{L \mid L \subset \mathbb{R}^4\}$, [27, 31, 33]

$$\mathcal{R}: C^\infty(\mathbb{C}^3) \to C^2(\mathbb{R}^4), \qquad \text{(VII. 4)}$$

with rule of correspondence

$$f(p, q, \zeta) \mid\!\to \int_{L \subset \mathbb{R}1} f(p, q, \zeta)\, d\zeta, \qquad \text{(VII. 5)}$$

which is an integral on a line of $\mathbb{C}^2$, of the bundle of lines seated in $\mathbb{F}$[10, 27].

Which is the behavior of these integrals on a complex Riemannian manifold?

Why is the special behavior of these integrals on a complex Riemannian manifold? Is possible establish criteria to the study of certain geometrical properties (such as curvature or torsion) of a complex Riemannian manifold **M**, based in integrability on a holomorphic complex bundle in a manifold through of their differentiable projections? Such criteria raise certain class of integrals calculated in sections of holomorphic bundle of M, in an analogous cohomology to the given by the space $H^{n-1,q}(\mathbb{C}^n/D, V)$?

Is feasible in this sense to obtain a generalization of cohomology on G-invariant orbits, belonging to holomorphic bundles seated in reductive homogeneous spaces of such form that are obtained integrals of generalized functional to the geometrical observables of the space **M**, and their orbits? Will establish this cohomology some theory of integration on certain complex spaces, parallel to the obtained by the integral cohomology on homogeneous bundles evaluating integrals in a direct and global via on the vector flux that conform them and define them?

The answer to first question save, clear this, the securing of harmonic functions ϕ, (it which no always is possible through of this path) is positive to the case of analytic functions and have that see with the structure a lo más Hermitiana of **M**, that can be establish constructing a fibration of **M** .

Then the resulting integrals are calculated on orbits of sections of the bundle TJ(**M**), the which is usually inherit the G-structure of J(**M**), the which in turn have the G-structure of **M**.

To the second question, is feasible to obtain such criteria by means of the use of tomography where the construction of the generalized functional required in the determination of geometrical invariants such as the curvature, requires the use of the Hermitian G-structure of the corresponding closed submanifolds used to the determination of the observable of the space in question. If we consider m = 1, and the Radon transform of the Dolbeault cohomology on functions $f \in D(\mathbf{M})$, (that is to say, with coefficients in a sheaf of functions in D(**M**)) then we can to induce anything commutative algebraic diagram of a vector bundle TD′(**M**), in a commutative algebraic diagram to a complex bundle $TD'(\mathbf{M} \otimes_{\mathbb{R}} \mathbb{C})$. The determination for example, of the curvature to this case is not trivial, and to it is required the use of certain compact components of a reductive homogeneous space whose orbital classes are holomorphic equivariant embeddings in the image of the endomorphism J, of the Hermitian structure of **M**. This it bears to the nullity of the anti-symmetric part of the tensor of curvature W_{ij}, (tensor de Weyl[5]), which is included with help of the Fröbenius theorem, since we consider the real analytic distributions of the bundle TJ(**M**). This establish a condition of integrability (VII. 1), that is a subtle pronouncement on the existence of a general integral to the equations of curvature of the Riemannian tensor to **M**, and that only depends of the Hermitian structure of the manifold **M**. Of this form all the differential geometry it could find their solution in an integral geometry, whose invariants could be

[5] tensor of Riemann = tensor of Ricci + tensor of Weyl. The tensor of Weyl represent the anti-symmetrical part of the tensor of curvature

fundament in Hermitian structures of a big collection of complex holomorphic manifolds. In this part result subtle the pronouncement of hypercomplex analysis on the fact of that hyperholomorphicity of several complex variables permits complex analyticity and real complex Riemannian manifolds, at least complex and to more Hermitians.

Some concrete examples we have in the identification of $\mathbb{H}$, with$\mathbb{C}^4$, which induce an application of complex lines that can be quaternionized. Indeed, be $\mathbb{H} = \{x = x_0 + ix_1 + jx_2 + kx_3 \mid x_m \in \mathbb{R}\}$, the algebra of the quaternions. The elements $x \in \mathbb{H}$, are explained of unique way as

$$x = z_1 + iz_2, \qquad \text{(VII. 6)}$$

considering the correspondences or equations of transformations

$$z_1 = x_1 + ix_2, \quad z_2 = x_3 + ix_4, \qquad \text{(VII. 7)}$$

which permit identify to $\mathbb{H}$, with $\mathbb{C}^2$. Also we can identify to $\mathbb{H}$, with $\mathbb{C}^4$, considering the space $\mathbf{E} \cong \mathbb{C}^4$, that is to say; considering the map

$$\mathbb{E} \to \mathbb{H}^2, \qquad \text{(VII. 8)}$$

with rule of correspondence

$$z \mapsto (w_0, w_1) := (z_0 + jz_1, z_2 + jz_3), \qquad \text{(VII. 9)}$$

that induce the application

$$\mathbb{P}^3(\mathbb{C}) \to \mathbb{P}(\mathbb{H}^2), \qquad \text{(VII. 10)}$$

that is to say, the corresponding quaternionic projective space of $\mathbb{P}^3(\mathbb{C})$, is $\mathbb{P}(\mathbb{H}^2)$. This space also is identified in twistor geometry like the corresponding twistor space of $\mathbb{H}$. Of this manage, the application (VII. 10), define a bundle with fiber $\mathbb{P}(\mathbb{C})$.

Thus exist a projection $\pi : \mathbb{P}^3(\mathbb{C}) \to S^4$, (the sphere of dimension four (which is a Riemannian manifold)), whose fibers are the real lines of $\mathbb{P}^3(\mathbb{C})$, that conforms the set of parameters of these straight lines in S^4. Observing that $\mathbb{P}(\mathbb{H}^2)$, have a recovering by charts each one is diffeomorphic to $\mathbb{R}^4$, that is that $\mathbb{P}(\mathbb{H}^2) \cong S^4$. Then considering open orbits in S^4, and the homogeneous holomorphic bundle of lines $O(k)$,on $\mathbb{P}^3(\mathbb{C})$, we have that the cohomological class $H^1(\pi^{-1}(\mathbf{x}), O(-2))$ $\forall \mathbf{x} \in S^4$, define a homogeneous bundle of lines on S^4. Then the subjacent orbital integrals in S^4, in the integral operator defined on this cohomology it is becomes in integrals of line on sections of the complex holomorphic bundle of lines $\mathcal{L}$, with cohomology $H^1(\mathbb{P}(\mathbb{H}^2), \mathcal{L})$, subjacent in the Penrose integral transform. The Hermitian structure in the corresponding manifold is invariant under these transformations, Where this the tomography? The tomography is realized on sections of the complex holomorphic bundle of lines of S^4, through of the lines of $\mathbb{P}^3(\mathbb{C})$. Is a form of re-write the vector Radon transform. If we consider the double fibration

$$\begin{array}{ccc} & \mathbf{F} & \\ \mu\swarrow & & \searrow\nu \\ \mathbf{P}^3(\mathbf{C}) & & G_{2,4}(\mathbf{C}) \\ \pi\searrow & & \swarrow\phi \\ & \mathbf{S}^4 & \end{array} \qquad \text{(VII. 11)}$$

The vector Radon transform that can be re-written be on the lines of S^4, as the integral operator

$$\mathcal{R}(\mathbf{u}, L) = \int_{L\subset S^4} \mathbf{u} \bullet d\mathbf{s}, \qquad \text{(VII. 12)}$$

finds their expression considering the flag manifold $\mathbb{F}$, given by the space $\{(L, P) | L \subset P \subset \mathbb{C}^4\}$, with the X-ray/John integral

$$\phi(P) = \int_{L\subset\mathbb{P}^3(\mathbb{C})} f, \qquad \text{(VII. 13)}$$

This integral comes of the application of the X-ray transform on lines of the projective bundle of lines in $\mathbb{P}^3(\mathbb{C})$, followed of the John transform. These integrals correspond to the orbital integrals of the Penrose transform that recovers harmonic functions of an open of S^4. Then in this sense, the cohomology $H^1(\pi^{-1}(\mathbf{x}), O(-2))$, not result very distinct of the cohomology $H^{n-1,q}(\mathbb{C}^n/D, V)$, save the fact of that the Hermitian structure deduced of the structure at least complex of **M**, results a relevant characteristic to the re-interpretation of the Radon integrals in complex holomorphic bundles of the manifold **M**. In this part is suggested the response to third question.

The fourth question is a difficult problem of decode and describe, and is a problem that can be resolve with help of the solution to the problem of the irreducible unitary representations in Lie groups to the space G/L, since is necessary the construction of a adequate correspondence between invariant bilinear forms and unitary representations of an analytic cohomology to some cases of finite dimension non-covered through this analysis.

In this sense, is necessary develop a way to modify the Dolbeault cohomology to produce minimal globalizations in major grade that the maximal globalizations calculated by Wong to the case of finite dimension. This establishes a relation of duality between the maximal globalizations calculated through of the Dolbeault cohomology and the minimal calculated for certain cohomology to define. How can identify the dual topological space of a cohomological space of Dolbeault on the non-compact complex manifold?

To respect results of great useful the holomorphic G-invariant vector bundle and their corresponding bundles of lines associated to the $(n, 0)$-forms on vector topological spaces. Then is obtained a version of the Dolbeault cohomology called topological cohomology of

Dolbeault and we construct representations of real reductive Lie groups G, beginning with the measurable complex flag manifold X = G/L, and using G-equivariant holomorphic bundle of lines on X, (conservation of the G-structure in submanifolds of X).

The hypercomplex analysis results useful only in the description of the unitary representations through of their realization and integral descriptions. The techniques of quaternions used describe some fields of some Lie algebras that are relevant and suggest the extension of those techniques to all the Lie algebras even of infinite dimension (Kac-Moody algebras, for example). Nevertheless, How it might be solved the problem of G-invariance on cycles and co-cycles of a complex Riemannian manifold of such lucky that it is not see affected the G-invariance of Hermitian structure to the symmetrical and non-symmetrical part of a complex Riemannian manifold, and be possible of be applying the methods of the hypercomplex analysis? Is possible establish a L^2-cohomology of integral operators based exclusively in the subjacent Hermitian structure of a complex manifold or at least complex?

To this last question, we consider the realization of holomorphic sections of certain homogeneous holomorphic bundles on $\mathbf{M}^+$, through an integral operator belonging to an integral L^2-cohomology. Consider the representation on L^2-holomorphic4-forms, where L^2,is defined with respect to invariant inner product

$$<\theta, \kappa> = \int_{M+} \theta \wedge \kappa, \qquad \text{(VII. 14)}$$

seated in $\mathbf{M}$, is the Flag bundle in $\mathbb{C}^4$,given by$\Im$. The pre-image of $\mathbf{M}^+$, under the natural projection is $\Im^{++-\!-}$, meaning that the Hermitian form have the indicated signatures (+, ++, ++– y ++––), when has been restricted to each part of the flag. This is one of six orbits of SU(2, 2), on $\Im$. Like is well knower, each orbit correspond to a different type of discrete series. This correspondence is realized making the cohomology of the appropriate bundle of lines on the several orbits. This process is analogous to discussed with anteriority to exhibit the equivalences of cohomological spaces on fibered in S^4,and $\mathbb{P}^3(\mathbb{C})$, but now is to the spaces $\mathbf{M}$, and $\Im$. Then to the orbit $\Im^{++-\!-}$, we note that the discrete series are $H^2(\Im^{++-\!-}, \Omega^6) \cong \Gamma(\mathbf{M}^+, \Omega^4)$. In this part result relevant the contribution of the generalized conform structure implicit on G, where maximal submanifolds of G, are horo-spheres, that is to say, bi-lateral components of maximal nilpotent subgroups. The transformation of a function to their integrals on horo-spheres play an important role in representation theory, such as is mentioned in this example. Their inversion is essential in the derivation of the Plancherel formula. Then, Is possible that SU(2, 2), can be represented on aL^2-cohomology based in integral operators on each one of their orbits? The answer is positive, and is possible through the twistor transform that is a SU(2, 2)-equivariant isomorphism, to know

$$\mathcal{T}: H^2(|\Im^{++-\!-}|, \Omega^6) \rightarrow H^3(|\Im - \Im^{++-\!-}|, \mathcal{O}), \qquad \text{(VII. 15)}$$

In the best of the cases, we could find an integral transform to each one of their orbits and for this way to find the complete discrete series. The true is that such isomorphism it is given in the level of minimal K-types of the representations [48].

This situation explain the importance of the G-structure of the manifold **M**, in the role on the determination of properties and invariants that can be appreciated on their orbits when this are K-invariants. But this can be studied for the via of the complex geometry of the vector G-bundles and the orbitalization of the reductive homogeneous spaces that are obtained in the process of reduction of G/H, which is not more than a consequence of the reduction of the group of holonomy of **M**.

We consider a K-invariant connection of reductive homogeneous space G/G_0, corresponding to a Stein manifold M_D, subjacent to the complex Riemannian manifold **M**, (that is to say, we consider a close and compact orbit of **M**). Then $G_0 = K$.

To it, we consider the cohomology of De Rham of the exterior subalgebras $\wedge(V)$,and$\wedge(V^*)$, of the vector bundle $E \rightarrow$ **M**, and we construct the K-invariant connection on the vector bundle $P \rightarrow$ **M**, (the vector G-bundle) that be an affine connection of **M**(2). For orbitalization

$$G/D \hookrightarrow G/P \hookrightarrow G/G_0, \quad G_0 = K, \tag{VII. 16}$$

we have that can to built a smooth embedding in J(**M**), of flag submanifolds $\mathbb{F}$, being G_0-orbits (is to say, K-orbits) in J(**M**). But this always is possible for the reduction of the holonomy group of M, and that M, be a connect and inner locally symmetric complex Riemannian manifold, that is to say, that the tensor of Nijenhuis satisfies on the corresponding sub-bundle J(**M**), of **M**, that $R_J^M(j) = 0$. This establishes the conditions of integrability to the symmetric part. Thus our sub-bundle have integrable structure and as the space considered is a manifold at least complex, one can to find complex submanifolds of J(**M**), to which the K-orbits are flag manifolds of the G-structure of **M**. From the point of view very particular of one study of the integral curvature [16, 17], this allow us to affirm that the contribution to the curvature on K-orbits only succeed in symmetric spaces (null Weyl tensor and null Ricci tensor), since the unique integrable complex submanifolds that can be realized like K-orbits in symmetric inner simply connect **M**, and of compact type are the flags G-orbits. In effect, for the arguments given to the second question, we observe that the use of the structure at least complex of the holomorphic bundle TJ(**M**), to the determination of fundamental 2-forms that evaluated in certain spaces result nulls, is a sufficient criteria to the integrability in **M**.

This condition establishes a criteria of *integrability* of the equations of curvature on the Riemannian manifold at least complex **M**. The class is of integrals that determine the functional that resolves the curvature equations belonging to integral operators L^2-cohomologyon submanifolds of **M**.

This geometrically is re-interpreted: the component of flat space in a Riemannian manifold at least complex result be the component with structure at least Hermitian, and given the local structure of Hilbert space that is subjacent in all Riemannian manifold, which can extend it to all the distributions, and in particular to the horizontal of the bundle TJ(**M**), result by the Fröbenius theorem, that the corresponding curvature tensor to the structure at least complex I, given by TJ(**M**) is vanished in said component, and not only that, also the Hermitian forms of the local structure of Hilbert space induces a symmetrical structure, and then in all component of flat space of **M**, are satisfied these conditions, transmitting the G-structure on the corresponding flag spaces (submanifolds of the corresponding complex holomorphic bundle TJ(**M**)), that in a adequate physical interpretation result be appropriates to the model and study of the phenomena to Max Planch lengths, writing and developing an integral operator cohomology based in integral operators whose integrals of contour are cohomological functional of energy states (cocycles).

Now well, Can these induce it the properties of the complex integral operators determined in the frame of a Hermitian structure of an Universe given for a complex Riemannian manifold to integral operators on a "microscopic" structure to the study of complex submanifolds more little that the flag manifolds, for example to null surfaces and null curves? Can be extended these operators to an integral operators cohomology isomorphic to the given by the resolution of an integral transform defined on homogeneous holomorphic bundles?

In this last point (on physic of particles) is induced to that the integral operators cohomology can be induced to a cohomology on diagrams of Feynman type. Then to this case arises the natural question of the cohomological classes to differential forms given ω, in the study on resolution of equations of vector fields ¿How re-emplace the integrals of contour

$$\int_{K} \omega, \tag{VII. 17}$$

by a conformal scheme (given, for example by $M_D \cong G/K$)) that helps to calculate these integrals of contour to these phenomena in the spaces of $\mathbb{F}$, and that not requires the evaluation of the vector field in orbital submanifolds like null curves, α–curves, minimal surfaces or null quadric?

For other side, we consider the problem of the integrability on complex Riemannian manifolds and their relation with the curvature tensor R^{jk}, to the conformal component of **M**.

Let G/L, be with L, a flag manifold. We consider the Radon transform on hyperplanes of arbitrary co-dimension of the Grassmannian $G_{n,m}$. Then to this particular case the integral operators cohomology on the complex flags is isomorphic to an integral operators cohomology on submanifolds of a complex maximal torus. Then the case of a symmetric connection to the space-time stay completely results, and can be computed the corresponding integrals, of this manage as study the space-time through of gauge fields on the which we can to calculate integrals on geodesics (for example, light geodesics to the determination of curvature of the space-time [18]). This defines an integral operators L^2-cohomology on orbits of the complex torus $\mathbb{T}$. The corresponding unitary representations are determined representations for quotients of the L^2-cohomology

$$\bigoplus_{\mu\in G} Hom_G(H_\mu, L^2(G/L)) \otimes H^s(G/L; \mathfrak{o}(\otimes_\mu)), \qquad \text{(VII. 18)}$$

If we consider $L = Z_G(\mathbb{T})$, the holonomic bundle that reduces the orbits of the homogeneous space G/K, to the orbits of the maximal torus in $\mathfrak{g}/\mathfrak{t}$,the which is a subspace of the Lie algebra of G/K, as can to see in the "orbitalization" of G/H

$$\begin{array}{c} J(\mathbb{R}^{2n}) \\ \cong \\ \mathfrak{g}/\mathfrak{t} \subset \mathfrak{g}/\mathfrak{p} \subset \mathfrak{g}/\mathfrak{so}(2n) \subset \mathfrak{g}/\mathfrak{h}, \end{array} \qquad \text{(VII. 19)}$$

then the quotiens are exhibited.

Here a flag manifold comes given like $\mathbb{F} = G/C(\mathbb{T})$, where $C(\mathbb{T})$, is a centralizator of the complex torus $\mathbb{T}$. Then the center $Z_G(\mathbb{T})$, induce a structure in $\mathbb{F}$, such that $Z_G = \{R_I^M(j) \in End(\wedge^2 T^*(\mathbf{M}) \otimes\otimes) \mid R_I^M(j) = 0\}$, and this consist of a finite number of connect components on each a of the G that act transitively.

Also any flag G-manifold[6] is realized like such orbit to some **M**. The requirement of the transitive action of G on the orbits is indispensable to the hypothesis of special isotropy that always is considered in the studies of the universe in the construction of an integral operators cohomology to the curvature of the space-time, since the integrals of curvature are defined on G-invariant orbits and will must be calculated by reduction of the corresponding holonomy group on K-invariant orbits.

The condition that is established with $R_I^M(j) = 0$,is the condition of integrability[7] of the symmetric connection that define the field equations in this case.

For filtration it can be possible to pass of the calculation of integrals on geodesic of the space-time to Feynman integral or integrals on strings considering the orbit of the $\mathfrak{g}$-filtration corresponding to the reductive homogeneous space of **M**. The process of reduction of the holonomy group of the structure of the vector G-bundle of **M** helps us to obtain reductive homogeneous spaces of G/H with inherit orbits the G-structure of the big space and therefore of **M**.

---

[6] Flag G-manifolds of **M**. G act transitively on such complex submanifolds.

[7] **Proposition (Burstall).** Be $j \in J(\mathbf{M})$. Then $G \bullet j$, is a submanifold at least complex of $J(\mathbf{M})$ on the which I, is integrable if and only if j, fall into of the null set of the Nijenhuis tensor $R_I^M(j)$.

Theorem VII. 1. (F. Bulnes). The K-invariance given by the G-structure S_G(**M**) of **M** complex and holomorphic is induced to each closed submanifold given by the flag manifolds of the corresponding vector holomorphic G-bundle. The integral operators cohomology given on such complex submanifolds is also equivalent to the integral cohomology on submanifolds of a complex maximum torus.

To demonstrate this result it is necessary to demonstrate some fundamental previous facts. For we consider it the following theorem due to (Burstall and Rawnsley, 1989), that is;

Theorem VII. 2. Let **M** = G/K, be a symmetrical internal simply related connect Riemanniana manifold and of compact type. Then

$$Z = \{R_I^M(j) \in End(\odot^2 T^*(M) \otimes \cdot \quad \cdot \;) \mid R_I^M(j) = 0\}, \qquad \text{(VII. 20)}$$

which consists of a related finite number of components on each one of those which G acts transitively. Any flag G-manifold is also carried out as such orbit for some **M**.

The requirement of the transitive action of G, on the orbits is for example, more important for the hypothesis of isotropy space in the construction of an integral operators cohomology for the curvature of the space-time, the integral of curvature should be determined since on G-invariants orbits and they will be calculated by reduction of the group of corresponding holonomy on K-invariants orbits.

Let us consider the space of classes G/C(T), which admits Kählerian complex invariant structures, that is to say; we can consider the G-structure K-invariant S_G(**M**), of **M**. To fix ideas we use the G-structure exactly as complex Kählerian of G/C(T). The complexified of the group G, given by G_C, acts transitively on G/C(T), for bi-holomorphism with parabolic groups as stabilizers. Reciprocally if $P \subset G_C$, is then a parabolic subgroup the action of G, on G_C/P, it has more than enough it is transitive and $G \cap P$, is the centralizing of a torus in G. For the infinitesimal situation (VII. 19) let F = G/C(T), be a flag manifold and let o the origin in F. Considers the decomposition of the Lie algebra $\mathfrak{g}_C$,

$$\mathfrak{g}_C = \mathfrak{h} \oplus \mathfrak{m}, \qquad \text{(VII. 21)}$$

with $\mathfrak{m} \cong T_o(F)$, and $\mathfrak{h}$, the Lie algebra of the stabilizer of o, in G. Then the complex invariant structure of F induces a decomposition ad$\mathfrak{h}$-invariant of $\mathfrak{m}_C$, in those (1, 0), and (0, 1), spaces

$$\mathfrak{m}_C = \mathfrak{m}^+ \oplus \mathfrak{m}^-, \qquad \text{(VII. 22)}$$

with $[\mathfrak{m}^+, \mathfrak{m}^+] \subset \mathfrak{m}^+$, for integrability. It can demonstrate himself that $\mathfrak{m}^+$, and $\mathfrak{m}^-$, are nilpotent subalgebras $\mathfrak{g}_C$, and in fact $\mathfrak{h}_C \oplus \mathfrak{m}^-$, is a parabolic algebra of $\mathfrak{g}_C$, with radical nilpotente $\mathfrak{m}^-$. If P, is then the corresponding parabolic subgroup of G_C/P, it is the stabilizer of o, and we obtain a bi-holomorphism among the complex space of classes G_C/P, and the flag manifold F.

Reciprocally, be P $\subset$Gc, a parabolic subgroup with Lie algebra $\mathfrak{p}$, and be $\mathfrak{n}$, the one conjugated of the radical nilpotent of $\mathfrak{p}$, (with regard to the real form $\mathfrak{g}$). Then H = G$\cap$ P, is the centralizing of a torus and we have the orthogonal decomposition (with regard to the Killing form, that is to say, the *ad*H-decomposition $\forall H\in\mathfrak{h}$):

$$\mathfrak{p} = \mathfrak{h}_C \oplus \underline{\mathfrak{n}}, \ \mathfrak{g}_\cdot = \mathfrak{h}_C \oplus \mathfrak{n} \oplus \underline{\mathfrak{n}}, \tag{VII. 23}$$

which defines a structure complex invariante G/H, has more than enough carrying out the bi-holomorphism with Gc/P. Considering an orbitalization of G/H, through the n-filtration of flag manifolds in Gc/P,

$$0 = \mathfrak{n}_{k+1} \subset \mathfrak{n}_k \subset \ldots \subset \mathfrak{n}_1 = \mathfrak{n}, \tag{VII. 24}$$

of $\mathfrak{n}$, defined for

$$\mathfrak{n}_1 = [\mathfrak{n}, \mathfrak{n}_{i-1}], \tag{VII. 25}$$

We realize orthogonalization this filtration using the form of Killing and we have that

$$\mathfrak{g}_i = \mathfrak{n}^{\perp}_{i+1} \cap \mathfrak{n}_i, \tag{VII. 26}$$

to i $\leq$1, and extended to a $\mathfrak{g}_C$, decomposition making correspond $\mathfrak{g}_0 = \mathfrak{h}_C = (\mathfrak{g} \cap \mathfrak{p})_C$, and $\mathfrak{g}_{-i} = \mathfrak{g}_i$, for i $\geq$1. Then

$$\mathfrak{g}_C = \sum \mathfrak{g}_i, \tag{VII. 27}$$

it is an orthogonal decomposition with

$$\mathfrak{p} = \sum\nolimits_{i \leq 0} \mathfrak{g}_i, \ \mathfrak{n} = \sum\nolimits_{i > 0} \mathfrak{g}_i, \tag{VII. 28}$$

The crucial property of this decomposition is that

$$[\mathfrak{g}_i, \mathfrak{g}_j] \subset \mathfrak{g}_{i+j}, \tag{VII. 29}$$

which can be proven demonstrating the existence of an element $\mathfrak{h}$, with the property that each iadξ, has eigenvalue$\sqrt{(-1)}$I, has more than enough $\mathfrak{g}_i$. This element ξ (necessarily only since $\mathfrak{g}$, is semi-simple) it was demonstrated that it exists in (Burstall and Rawnsley, 1989), which called it canonical element of $\mathfrak{p}$. Since *ad*ξ has eigenvalues in $\sqrt{(-1)}\cdot$, Ad exp$\pi\xi$, is an involution of $\mathfrak{g}$, the one which exponencing to obtain an internal involution τ_ξ, of G, and therefore an internal symmetrical space G/K, where K = $(G^{\tau\xi})_0$. Clearly, K, has the Lie algebra given for

$$\mathfrak{k} = \mathfrak{g} \cap \sum \mathfrak{g}_i, \tag{VII. 30}$$

and that H contains, of where we obtain the homogeneous fibration

$$G/H \rightarrow G/K, \qquad (VII. 31)$$

of our flag manifold on our symmetrical space. This will be important to establish an isomorphism among the cohomology of the integrals on G/H, and the corresponding integral submanifolds that are had in the ker{$\underline{\partial}$-equations}, that like it has been said, they only exist as integral of those $\underline{\partial}$-equations in **M**, (**M**, integrable) if $R_I^M(j) = 0$. The filtration is also essentially only. The only ambiguity in the prescription is that the distinct points in the symmetrical space can have the same stabilizer K (that is to say, antipodal points in a sphere). Nevertheless the number of such points is finite and we can give a finite number of such fibrations. These fibrations will call them canonical of F. Then it is necessary to enunciate the following result (Burstall and Rawnsley, 1989):

Theorem VII. 3. Let $F = G_{\mathbb{C}}/P$, be a flag manifold. Then an only internal symmetrical G-space always exists **M**, associated to F, with a number of homogeneous fibrations $F \rightarrow M$.

For the demonstration of this result, let us consider a conjugation class $\mathfrak{p}$, of $\mathfrak{g}c$, and the election in a real way of $\mathfrak{g}$. Now what it is necessary to see is that each flag manifold is a fiber on each internal symmetrical space. Reciprocally, this it is the road to demonstrate that each space symmetrical intern is the objective of the canonical fibrations of at least a flag manifold. This finally will give an idea of how the geometry of the complex holomorphic bundle is J(**M**).

Be

$$p: F \rightarrow \mathbf{M},$$

a canonical fibration. For construction, the fibers of p are complex submanifolds of F, and this facilitates us to define a fiber map

$$i_p: F \rightarrow J(\mathbf{M}), \qquad (VII. 32)$$

as it continues: In $f \in F$, has the orthogonal decomposition of T_fF, in horizontal and vertical subspaces, both of which are then low invariants the complex structure of F. Then dp restrict giving an isomorphism of the horizontal part with $T_{p(f)}\mathbf{M}$, and therefore it induces a structure at most Hermitiana $T_{p(f)}\mathbf{M}$, it has more than enough, this is; $i_{p(f)}(f) \in J_{p(f)}\mathbf{M}$. Such a construction is possible provided we have a Riemannian submersion of a Hermitian manifold with submanifolds like fibers. For this case we consider:

Proposition VII. 1. $i_p: F \rightarrow J(\mathbf{M})$, it is a fitting holomorphic G-equivariant.

It proves: (Burstall, 1987).

This implies that $i_p(F)$, it is at most a complex submanifold of J(**M**), on which **I**, is integrable. For a corollary of the Burstall theorem of the proposition II. 1., we have:

Corollary VII. 1. $i_P(F)$ it is a G-orbit in $Z \subset J(\mathbf{M})$. It proves in, (Burstall, 1987).

In particular this guarantees that the tensor space of Z is I don't empty. The reciprocal of this corollary one is also true.

The following result due to Rawnsley determines which the fitting is given by the proposition II. 1., and the theorem II. 3:

Theorem VII. 4. If $j \in Z \subset J(\mathbf{M})$ then $G \bullet j$, is canonically a flag fibreded manifold **M**. It has more than enough in fact, $G \bullet j = i_P(F)$ for some fibration $p: F \to \mathbf{M}$ of a flag manifold F.

Their demonstration is continued of the observation that in $\pi(j)$, we have the symmetrical decomposition

$$\mathfrak{g} = \mathfrak{k} \oplus \mathfrak{q}, \tag{VII. 33}$$

with $\mathfrak{q} \cong T_{p(f)}\mathbf{M}$. If $\mathfrak{q}$, is the (0, 1)-space for j, then

$$[\mathfrak{q}^-, \mathfrak{q}^-] \oplus \mathfrak{q}^-, \tag{VII. 34}$$

it is the nilpotent radical of the parabolic subalgebra p. Then it is demonstrated that $G \bullet j$, is equivariantly bi-holomorphic to the corresponding flag manifold such G_C/P, and like it is described in the theorem. II. 4. For more details see (Burstall and Rawnsley, 1989).

Now complete the demonstration of the theorem. II. 2. We have to see that each canonical fibration of a flag manifold will give to a G-orbit in Z, for some internal symmetrical G-space **M**, and that all such orbits come from the same procedure. But fixed G, stops, it exist alone a finite number of flag manifolds of bi-holomorphism of this type. These are in biyective correspondence with the conjugated classes of parabolic subalgebras of $\mathfrak{g}_C$, and each flag manifold admits a finite number of canonical fibrations. Then Z, is made up of a finite number of G-orbits all of which are closed and the theorem II. 2., it is continued. Then since each one of these G-orbits exists like a K-orbit of the space of classes G/K, with tensor of curvature of null Nijenhuis then each flag submanifold is a K-orbit of the vector and holomorphic G-bundle of the 2n-dimensional irreducible symmetrical Riemanniana manifold $J(\mathbf{M})$, reason why the theorem II. 1., is continued.

Chapter VIII

Orbital Integrals on Cuspidal Forms

VIII. 1. Introduction

To can to study the cuspidal forms and determine their orbital integrals on whose cuspidal forms is necessary to do the use of the orbital classes of the group SL(2, ℝ), and of their corresponding Lie algebra $\mathfrak{sl}(2, \mathbb{R})$. To them, we identify to a Lie group L, locally isomorphic to SL(2, ℝ), and whose Lie algebra corresponding it is identify isomorphically with $\mathfrak{sl}(2, \mathbb{R})$. The orbital integrals as obtained will be orbital integrals of the minimal parabolic subgroup in G. Applying the Harish-Chandra to the obtained functions of such orbital integrals we will obtain the orbital integrals on cuspidal forms.

We begin this section with some computes on the group SL(2, ℝ). All the results it is will must to Harish-Chandra.

Let L, be a locally connect Lie group isomorphic to SL(2, ℝ). We identify the corresponding Lie algebra of L, with the Lie algebra $\mathfrak{sl}(2, \mathbb{R})$.

VIII. 2. Exposition

Let T, be A^0, and N, connect subgroups of G, whose corresponding Lie algebra are the spaces

$$\mathbb{R}h = \{h \in \mathfrak{t} \mid h = \begin{bmatrix} 0 & 1 \\ -1 & 0 \end{bmatrix}\}, \quad \mathbb{R}H = \{H \in \mathfrak{a}^0 \mid H = \begin{bmatrix} 1 & 0 \\ 0 & -1 \end{bmatrix}\}, \qquad \text{(VIII. 1)}$$

and

$$\mathbb{R}X = \{X \in \mathfrak{m} \mid X = \begin{bmatrix} 0 & 1 \\ 0 & 0 \end{bmatrix}]\}, \qquad \text{(VIII. 2)}$$

Let A, be a Cartan subgroup of G, corresponding to Lie algebra $\mathfrak{a} = \mathfrak{a}^0 \oplus \mathfrak{t}$. Be $\forall \theta \in \mathbb{R}$, the map

$$t : \mathbb{R} \rightarrow T, \qquad \text{(VIII. 3)}$$

whose rule of correspondence is

$$\theta \mid\!\rightarrow \exp(\pi\theta h), \quad \forall\ h \in \mathfrak{t}, \ \ \theta \in \mathbb{R}, \qquad \text{(VIII. 4)}$$

whose rule explicit is

$$t(\theta) = \exp(\pi\theta h), \qquad \text{(VIII. 5)}$$

If $f \in \mathscr{C}(G)$, then $F_f^T(t(\theta)) = F_f^T(\theta)$. In effect, by definition

$$F_f^T(t(\theta)) = \Delta(t(\theta))\int_G f(gt(\theta)g^{-1})dg = \Delta(\exp(\pi\theta h))\int_G f(g\exp(\pi\theta h)g^{-1})dg, \qquad \text{(VIII. 6)}$$

But

$$t(\theta) = \exp\pi\theta\begin{bmatrix} 0 & 1 \\ -1 & 0 \end{bmatrix} = \exp\begin{bmatrix} 0 & \pi\theta \\ -\pi\theta & 0 \end{bmatrix} = \begin{bmatrix} 0 & e^{\pi\theta} \\ -e^{-\pi\theta} & 0 \end{bmatrix}, \qquad \text{(VIII. 7)}$$

Then

$$\Delta\begin{bmatrix} 0 & e^{\pi\theta} \\ -e^{-\pi\theta} & 0 \end{bmatrix}\int_G f(g\begin{bmatrix} 0 & e^{\pi\theta} \\ -e^{-\pi\theta} & 0 \end{bmatrix}g^{-1})dg \quad \int_G f(gg^{-1})dg$$

$$= 2\text{sen}\ \pi\theta\int_G f(g\pi\begin{bmatrix} 0 & e^{\theta} \\ -e^{-\theta} & 0 \end{bmatrix}, g^{-1})dg = F_f(\theta),$$

Note that $T'' = T' = \{t(\theta)\,|\,\theta\in\mathbb{R}\backslash\mathbb{Z}\}$. In effect, for one side $T'' = T \cap K''$, with

$$K'' = \{k\in K\,|\,(I - Ad(k)|_{\mathfrak{p}}) \neq 0\},$$

Since K'', is maximal in $G''[K]$, of G, we have that $T \cap K'' = \{t\in T|(I - Ad(t)|_{\mathfrak{t}''}) \neq 0\}$. But since to each K'' exist $k_1\in\mathbb{Z}$, such that $(I - ad(H)|_{\mathfrak{t}''})^{k_1} = 0$, and given that $\mathfrak{g} = \mathfrak{t} \oplus \mathfrak{p} \oplus \mathfrak{a}$, then exist $k_2\in\mathbb{R}\backslash\mathbb{Z}$, such that $(I - ad(H)|_{\mathfrak{p}})^{k_2} = 0$. Thus, and in particular to the classes $\mathbb{R}h$, in $\mathbf{t}$,

$$T'' = T' = \{t(\theta)\,|\,\theta\in\mathbb{R}\backslash\mathbb{Z}\}. \qquad \text{(VIII. 8)}$$

A direct computing, using the formulas in 7. 4. 3 and 7. 4. 4., give

1. $F_f(\theta) = 2i\text{sen}\ \pi\theta\int_0^\infty \text{senh}(2t)f(\exp(\theta\pi\begin{bmatrix} 0 & e^{2t} \\ -e^{-2t} & 0 \end{bmatrix})) dt,$

Let $u = |\pi\theta| \cosh(2t)$, be. Then we have to $\theta \neq 0$,

2. $F_f(\theta) = 2i(\text{sen}\ \pi\theta/|\pi\theta|)\int_{\pi\theta}^\infty f(\exp(\text{sgn}\theta\begin{bmatrix} 0 & z(\theta,u) \\ z(\theta,-u) & 0 \end{bmatrix})) du,$

with $z(\theta, u) = u + (u^2 - (\pi\theta)^2)^{1/2}$. The two values (mod $2\mathbb{Z}$), to which we have singularities of type "jump" are $\theta = 0$, or 1. The before formula show that this is not a singularity "jump" to $\theta = 1$. We concentrate to case $\theta = 0$. Likewise, (2) implies that:

$$(3)\lim_{\theta\longrightarrow 0^+} F_f(\theta) = 2i\int_0^{\infty} f(\exp(2Xu))du, \qquad and \qquad \lim_{\theta\longrightarrow 0^-} F_f(\theta) = 2i\int_0^{\infty} f(\exp(-2Xu))du.$$

This implies that:

$$(4)\lim_{\theta\longrightarrow 0^+} F_f(\theta) - \lim_{\theta\longrightarrow 0^-} F_f(\theta) = i\int_{-\infty}^{\infty} f(\exp(Xu))du.$$

If we derive the formula (2), we have

$$\left(\frac{d}{d\theta}\right)F_f(\theta) = \frac{\theta(\pi\theta\cos\pi\theta - \operatorname{sen}\pi\theta)}{\pi\theta^2\operatorname{sen}\pi\theta}F_f(\theta) - 2\pi i(\frac{\operatorname{sen}\pi\theta}{\pi\theta})f(\exp\pi\theta h) + E(\theta))) \qquad \text{(VIII. 9)}$$

with $\lim_{\theta\to 0}E(\theta) = 0$. Thus it is have that

$$(5)\lim_{\theta\longrightarrow 0}\left(\frac{d}{d\theta}\right)F_f(\theta) = 2\pi i f(1).$$

Now re-define (4), and (5), in terms of integrals on the subgroup of Cartan A of G. To it, you consider the endomorphism $H_f(t) = F_f^A(\exp(tH))$. Then by (III.79) and (III. 80), we affirm to this case that:

$$(6)H_f(t) = e^t\int_{-\infty}^{\infty} f(\exp tH\exp Xx))dx.$$

of where thus it is conclude that

$$(7)\lim_{\theta\longrightarrow 0^+} F_f(\theta) - \lim_{\theta\longrightarrow 0^-} F_f(\theta) = i\lim_{t\longrightarrow 0} H_f(t).$$

But the definition of F_f, implies that $F_f(a) = F_f(a^{-1})$. Thus

$$\lim_{t\longrightarrow 0}(d/dt)H_f(t) = 0, \qquad \text{(VIII. 10)}$$

then

$$(8)\lim_{\theta\longrightarrow 0^+}\left(\frac{d}{d\theta}\right)F_f(\theta) - \lim_{\theta\longrightarrow 0^-}\left(\frac{d}{d\theta}\right)F_f(\theta) = 0 = \lim_{t\longrightarrow 0}\left(\frac{d}{dt}\right)H_f(t).$$

Be $C\in Z(\mathfrak{L}_C)$, such that $\gamma_T(C) = -(d^2/d\theta^2)$, and $\gamma_A(C) = d^2/dt^2$. Here γ_J, is the isomorphism of Harish- Chandra associated with the subgroup of Cartan J. C, (is save a scalar multiple and subtraction of a scalar) the Casimir operator of the algebra $\mathfrak{L}$. The Casimir operator, and their analogous to F^A, combined with (7), and (8), imply that

$$(9)\ \lim_{\theta\longrightarrow 0^+}\left(\frac{d}{d\theta}\right)^k F_f(\theta) - \lim_{\theta\longrightarrow 0^-}\left(\frac{d}{d\theta}\right)^k F_f(\theta) = 0 = i^{k+1}\lim_{t\longrightarrow 0}\left(\frac{d}{dt}\right)^k H_f(t).$$

We assume that $G = G^0$, where

$${}^0G = \{g\in G \mid Ad(g)^k X = 0,\ \forall\ X\in Ж(G)\},$$

where $Ж(G) = \{\chi\in Hom(G, \mathbb{R}^*)|\chi$ is continuous$\}$. We assume that G is of inner type ($Ad(G)\subset G$). Be $A\subset H$. Then

$$F_f^H(h) = \Delta(h)\int_G f(ghg^{-1})dg,$$

If $f\in \mathscr{C}(G)$, and if $a\in A$, then $R(a)f|_{0G} = u\in \mathscr{C}({}^0G)$. In consequence, $F_f^H(ha) = F_u^{H\cap 0G}(h)$ to $h\in H\cap{}^0G$.

This will be of utility to transfer the results of the case $G = {}^0G$, to a more general situation.

Theorem VIII. 2. 1. Be $f\in \mathscr{C}(G)$, and we assume that $F_f^H = 0$, to each subgroup of Cartan of G, that is not fundamental. If H, is fundamental then F_f^H, extends to a smooth function on H.

*Proof.*In the demonstration of the theorem is necessary to use orbital integrals on the different real reductive groups. If L, is a reductive group and if J, is a subgroup of Cartan of L, then we define

$${}^LF_f^J = F_f, \tag{VIII. 11}$$

This notation will can to help to identify to real reductive group on which the integration have happen.

That's right; we demonstrate the result by induction on the dimension of G.

i. If $\dim G = 0$, or 1, then $G = H$, is the unique subgroup of Cartan. Then, if $f\in \mathscr{C}(G)$,

$$F_f^H(h) = \Delta_H(h)\int_{H/H} f(h)dhH = \Delta_H(h)f(k), \tag{VIII. 12}$$

But $\Delta_H(h) = \pm\Delta(h)$, $\forall\ h\in H$. If $H = G$, then $\Delta_H(h) = \pm\Delta(h) = \pm 1$, of where

$$F_f^H(h) = f(h),\quad \forall\ h\in H.$$

ii. Suppose that the result is valid to the cuspidal form ${}^LF_f^J$, with J, a subgroup of Cartan of the real reductive group L, with $0 \le \dim L \le \dim G$. Is necessary demonstrate the result on a cuspidal form ${}^{0G}F_f^J$,of ${}^0G$, a real reductive group whose subgroup of Cartan is J, $f\in \mathscr{C}(G)$, such that

$$0 \le \dim L \le \dim {}^0G \le \dim G, \tag{VIII. 13}$$

In effect, if $G \neq {}^0G$, then dim ${}^0G<$ dim G, for be ${}^0G$, a connect subgroup of G (identity component of G). Be A, a splitting component of G. If J, is a subalgebra of Cartan of ${}^0G$, then JA, is a subalgebra of Cartan of G, and each subalgebra of Cartan of G, is of this form. That's right, only is necessary consider the cuspidal form

$$F_f^{FA}(ja) = {}^{0G}F_f^J(j) = F_f^J(j),$$

and to use the Harish-Chandra transform $f^P(ja) = u_a(j)$, to realize a right translation of $f\in \mathscr{C}(G)$, and of that manage to obtain a extension of F_f^J, on L, like smooth function in L, (If $f\in \mathscr{C}(G)$, and if $a\in A$, then $R(a)f|_{0G} = u\in \mathscr{C}(G)$). Thus

$$F_f^{FA}(ja) = F_{u_a}^{J\cap 0G}(j) = F_f^J(j), \quad \forall\ j\in J\cap {}^0G, \qquad \text{(VIII. 14)}$$

Then (VIII. 13), follows. Thus F_f^H, it is can extend like smooth function to H, a subgroup of Cartan of G.

Now we assume that $G = {}^0G$. Also we assume in G, that H, is a non-compact fundamental subgroup of Cartan of G. Then we can assume that $H = H_F$, and P_F, is proper in G. We consider $Q = P_F$, $L = {}^0M_F$, and $T = T_F$. If J, is a subgroup of Cartan of L, then JA, is a subgroup of Cartan of G, and if $f\in \mathscr{C}(G)$,

$$F_f^{JA}(ja) = {}^LF^J_{R(a)fQ}(j), \quad \forall\ j\in J,\ a\in A_F.$$

Thus the induction hypothesis to this non-compact case also prevalence.

If T is a compact subgroup of Cartan of G, then

$$F_f^T(t) = F_f(t), \quad \forall\ t\in T,\ f\in \mathscr{C}(G).$$

Let $\Phi_n = \{\alpha\in\Phi^+|\ (\mathfrak{g}_C)_\alpha\subset\mathfrak{p}_C\}$. Let $\sigma X = \underline{X} = conjX$, $\forall\ X\in\mathfrak{g}_C$, respect to $\mathfrak{g}$. Then $\sigma(\mathfrak{g}_C)_\alpha = (\mathfrak{g}_C)_{-\alpha}$. Let $\alpha\in\Phi_n$, $Z\in(\mathfrak{g}_C)_\alpha$, and $W = \sigma Z = conjZ = \underline{Z}$. If $Z \neq 0$, then $Z + W \neq 0$, in $\mathfrak{p}$ (just not $\mathfrak{p}_C$) since $Z + \underline{Z} = 2X\in\mathfrak{p}$.

Consider the map

$$[,] : (\mathfrak{g}_C)_\alpha \times (\mathfrak{g}_C)_{-\alpha} \rightarrow [(\mathfrak{g}_C)_\alpha, (\mathfrak{g}_C)_{-\alpha}], \qquad \text{(VIII. 15)}$$

whose rule of correspondence be

$$(Z, W) \mapsto [Z, W], \qquad \text{(VIII. 16)}$$

followed of the map

$$\alpha: [(\mathfrak{g}_C)_\alpha, (\mathfrak{g}_C)_{-\alpha}] \rightarrow \mathbb{R}, \qquad \text{(VIII. 17)}$$

whose rule of correspondence is:

$$[Z, W] \mapsto \alpha([Z, W]), \tag{VIII. 18}$$

we have that $\alpha([Z, W]) = \alpha(ZW - WZ)$, and give the normalization $Z/\|Z\|^2$, we have

$$\frac{\alpha(ZW - WZ)}{\|Z\|^2} = \frac{\alpha ZW - \alpha WZ}{\|Z\|^2} = \frac{(\alpha\|Z\|^2 - \alpha\|\underline{Z}\|^2)}{\|Z\|^2} = \frac{(\|Z\|^2 + \bar{\alpha}\|Z\|^2)}{\|Z\|^2} = \frac{2\|Z\|^2}{\|Z\|^2} = 2,$$

given that $\|Z\|^2 = \|\underline{Z}\|^2$. Thus $\alpha([Z, W]) = 2$. Calling $H = Z + W$, and $h = -i[Z, W]$, and $X = (1/2)([Z, W] + i(Z - W))$. It is can verify that H, h, and X, complies with the same relations of commutation like the given to the elements of $\mathfrak{sl}(2, \mathbb{R})$.

Let $\mathfrak{l}^\alpha = \mathbb{R}H + \mathbb{R}X + \mathfrak{t}$. Then $[\mathfrak{l}^\alpha, \mathfrak{l}^\alpha] \cong \mathfrak{sl}(2, \mathbb{R})$. Let the orbits of T, $T_\alpha = \{t \in T \mid t^\alpha = 1\}$, and

$T'_\alpha = \{t \in T_\alpha \mid t^\beta \neq 1, \forall \beta \in \Phi^\alpha - \{\alpha\}\}$, then the space $T'_\alpha \exp(\mathbb{R}H)$, is open in T. That' right, be L^α, the connect subgroup of G, with Lie subalgebra $\mathfrak{l}^\alpha$. Then T_α, is the center of L^α. Let $k_\alpha(\theta) = \exp\theta\pi h$. If $t = uk_\alpha(\theta) \in T''$, and if $f \in C_c(G)$, then

iii
$$\int_G f(xtx^{-1}) = \int_{G/L^\alpha}\int_{L^\alpha} f(guxk_\alpha(\theta)x^{-1}g^{-1})dxdg,$$

Considering $\Delta_\alpha(t) = t^{\rho - \alpha/2}\Pi_{\beta \in \Phi - \{\alpha\}}(1 - t^{-\beta})$, then

iv
$$\Delta(uk_\alpha(\theta)) = 2i\Delta_\alpha(uk_\alpha(\theta))\text{sen}(\pi\theta),$$

Let the cuspidal form on L^α,

$$R_f(g,u,\theta) = \Delta_\alpha(uk_\alpha(\theta))\text{sen}(\pi\theta)\int_{G/L^\alpha} R_f(g,u,\theta)dgL^\alpha,$$

Then on the classes in G/L^α, we can consider it, cuspidal form

$$F_f(uk_\alpha(\theta)) = \int_{G/L\alpha} R_f(g, u, \theta)dgL^\alpha,$$

Let $f \in C_c^\infty(G)$. Let $u \in T'_\alpha$. We fix $p \in U(\mathfrak{t}_C)_\alpha$. Note that if $|\theta| < \varepsilon$, with ε, enough little and such that $|\theta| > 0$, then $uk_\alpha(\theta) \in T'$. Furthermore, writing for the derivation rule of the cuspidal form $F_f(uk_2(\theta))$, that

$$ph^k F_f(uk_\alpha(\theta)) =$$
$$= \Sigma\binom{k}{j}\left(i\frac{d}{d\theta}\right)^{k-j} p\Delta_\alpha(uk_\alpha(\theta))\left(i\frac{d}{d\theta}\right)^j \text{sen}\pi\theta\int_{G/L} R_f(g,u,\theta)dgL^\alpha,$$

Let J, be the centralizator of G, of $T_\alpha\exp(\mathbb{R}H)$. Then using (9), save a constant, we have

$$\lim_{\theta\to 0^+} ph^k F_f^{\ T}(uk_\alpha(\theta)) - \lim_{\theta\to 0^-} ph^k F_f^{\ T}(uk_\alpha(\theta))$$
$$= p\Sigma\begin{pmatrix} k \\ j \end{pmatrix}\left(i\frac{d}{d\theta}\right)^{k-j}(\Delta_\alpha(uk_\alpha(\theta)))\Delta_\alpha(u)^{-1}\lim_{t\to 0}\left(\frac{d}{dt}\right)^j F_f^{\ T}(u\exp H),$$

Since the both sides of (iv), are continuous on $\mathscr{C}(G)$, it is $f\in\mathscr{C}(G)$. This is the condition of jump mentioned to begin of the demonstration of theorem. Then the formula (v), implies that if $F_f^H = 0$, $\forall$ the non-fundamental subgroups of Cartan H, of G, then F_f, is smooth in a neighborhood of each $t\in T'_\alpha$, $\forall\alpha\in\Phi_n^+$, is to say, to all point in the space $T'_\alpha\exp\mathbb{R}H$. Now well, suppose that $\alpha\in\Phi_k^+$. Be

$$\mathfrak{l}^\alpha = \mathfrak{g}\cap(\mathfrak{t}_\mathbb{C} + (\mathfrak{g}_\mathbb{C})_\alpha + (\mathfrak{g}_\mathbb{C})_{-\alpha}), \qquad \text{(VIII. 19)}$$

Let L^α, be the connect subgroup of G, corresponding to $\mathfrak{l}^\alpha$. Then L^α, is compact. By the cuspidal lemma, the differences given in (v), are not jumps, are cusped in this case.

Then it is have demonstrate that: If $F_f^H = 0$, to all the subgroups of Cartan non-fundamentals of G, then the cuspidal form F_f^H, is smooth in a neighborhood of each $t\in T$, such that $t^\alpha = 1$, to at least a $\alpha\in\Phi^+$. Then

$$T'_\alpha\exp(\mathbb{R}h)\cap T_\alpha\exp(\mathbb{R}H) = T'_\alpha\exp(\mathbb{R}h),$$

is open in T, and since pf, it is extends like a smooth map on T'_α, $\forall$ $p\in U(\mathfrak{t}_\mathbb{C})$, then f, it is extends like a smooth function on T. ∎

One direct application is:

Corollary VIII. 2. 1. Be $f\in\mathscr{C}(G)$, a cuspidal form. If H, is a subgroup of Cartan of G, whose compact module is not the centre of G, then $F_f^H = 0$. If H, is compact module that is centre of G, then F_f^H, it is extend to a smooth function on H.

Proof. See [1].

Chapter IX

Some Applications in the Study of Harmonic Analysis

IX. 1. Introduction

We remember that a cuspidal form cans to be defined to be defined by means of the Harish-Chandra transform as those function $f\in \mathscr{C}(G)$, such that $(L(x)R(x)f)^P = 0$, with P a proper parabolic subgroup of G, and $\forall$ x, y$\in$G.

But due to that the space of the cuspidal forms E_f, can to be identified like the closure space $Cl(\mathscr{C}(G))$, of the matrix coefficients K-finites of the discrete series then by the theorem on closure of the space $\mathscr{C}(G)$ [39], can to be assumed that

$$E_f = \{f\in \mathscr{C}(G) \mid \dim Z_G(\mathfrak{g}) < \infty\} \cong {}^0\mathscr{C}(G), \quad \text{(IX. 1.1)}$$

where

$${}^0\mathscr{C}(G) = \{f\in \mathscr{C}(G) \mid (L(x)R(x)f)^P = 0,\ \forall\ x, y\in G \text{ and } P = {}^0MN \subset G\}, \quad \text{(IX. 1.2)}$$

Likewise, if we consider all the compact Cartan subgroups T, of G, result some interesant applicable properties of harmonic type to cuspidal forms space ${}^0\mathscr{C}(G)$.

Considering that all the Cartan subgroups of G, were unidimensionals, we can identify to Lie algebra $\mathfrak{g}$, of G, as the corresponding algebras of the subgroups SL(2, $\mathbb{R}$), and SU(2). For it is enough to consider the integrals of cuspidal forms on the orbits T'_α, and T_α, of T, $\forall\alpha\in\Phi^+_k$.

For the case of compact Cartan subgroups of G, whose dimension to be major that 1, it is possible to apply the Peter- Weyl theorem and the orthogonality relations of Schür and obtain an orbital algebra analogous to space

$$l^\alpha = g \cap (t_C + (g_C)_\alpha + (g_C)_{-\alpha}), \quad \text{(IX. 1.3)}$$

We can thus to assum that the Cartan subgroups of G, could be at least 2-dimensional.

IX. 2. Harmonic analysis of cuspidal forms

Let G, be a real reductive group such that $Ad(G) \subseteq G$, and $G = {}^0G$. Let's visualize to symmetric algebra $S(\mathfrak{g}_C)$, as the algebra of differential operators with constant coefficients on

$\mathfrak{g}$. Let $\theta \in \mathbb{R}$, be such that $t(\theta) = \exp(\pi\theta h)$, $\forall \theta h \in \mathbb{R}H$. If $f \in \mathscr{C}(G)$, then $F_f^T(f(\theta)) = F_f(\theta)$. Let $q = \dim \mathfrak{t}$, and $p = \dim \mathfrak{p}$.

Let ${}^0\mathscr{C}(G)$, like was defined in (IX. 1. 2). If G, have a compact Cartan subgroup T, and if Φ^+, is a positive roots system to $\Phi(\mathfrak{g}_C, \mathfrak{t}_C)$, then

$$\varpi = \prod_{\alpha \in \Phi^+} H_\alpha \in U(\mathfrak{t}_C), \qquad \text{(IX. 2.1)}$$

Theorem IX. 2.1. If G, have non-compact Cartan subgroups then ${}^0\mathscr{C}(G) = \{0\}$. If T, is compact Cartan subgroup of G, then there is a non-nule constant C_G, such that if $f \in {}^0\mathscr{C}(G)$, then

$$\varpi F_f^T(1) = C_G f(1), \qquad \text{(IX. 2.2)}$$

The result is a special case of a more general theorem of Harish-Chandra, which establish a similar formula $\forall$ $f \in \mathscr{C}(G)$, replacing T, for fundamental Cartan subgroups. Since, we come to generalize on the compact Cartan subgroups of G, then we demonstrate first that ${}^0\mathscr{C}(G) = \{0\}$, if G, have non-compact Cartan subgroups. For it, we have that consider some important facts on the formulas of the theorem 6.7.1, to this concrete case and two lemmas that were demonstrated in their time.

Let $X_1, \ldots, X_n$, be a pseudo-base of $\mathfrak{g}$, relative to B (that is to say $B(X_j, X_k) = \delta_{j,k}$) such that $\mathfrak{g} \cong \mathbb{R}^n$. Of fact, through the diagram

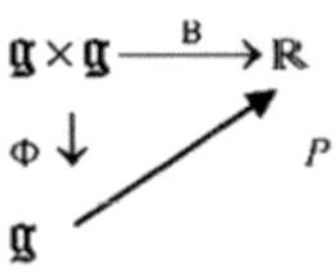

(IX. 2.3)

we can deduce the composition map

$$B = \Phi \circ P : \mathfrak{g} \times \mathfrak{g} \to \mathfrak{g} \to \mathbb{R}, \qquad \text{(IX. 2.4)}$$

whose rule of correspondence is

$$B : (X, X) \mapsto P(X), \qquad \text{(IX. 2.5)}$$

Then $\varpi = \sum X_j Y_j$, with $X_j, Y_j \in \{X_k\}$, of $\mathfrak{g}$, relative to B. Since $n \geq 2$, then $n = p + q$, $\forall$ $p, q \in \mathbb{Z}^+$, and given that $\mathfrak{g} \cong \mathbb{R}^n$, then

$$\mathbb{R}^n = \mathbb{R}^p \oplus \mathbb{R}^q = \mathfrak{p} \oplus \mathfrak{t}, \qquad \text{(IX. 2.6)}$$

Then we define to $p, q \geq 1$, the cuspidal form $F_{p,q} = F_{\mathfrak{g}} = F$, such if $F_{p,q}$, is integrable then there is a not null constant $C_{p,q}$, such that

$$\int_{\mathfrak{g}_e} F_{p,q}(x)\overline{\varpi}^{[n/2]}f(x)dx = C_{p,q}f(0), \qquad \text{(IX. 2.7)}$$

Since that $F = F_{p,q}$, $\forall$ p, q ≥ 1, is G-invariant then $F(Ad(g)X) = F(X)$, $\forall$ $g \in G$, and $X \in \mathfrak{g}$. If we consider a system of roots $\Phi_j \subset \Phi^+$, relative to $\Phi(\mathfrak{g}_C, (\mathfrak{h}_j)_C)$, we can define

$$\pi_j(h) = \prod_{\alpha \in \Phi_j} \alpha(h), \qquad \text{(IX. 2.8)}$$

$\forall$ $h \in \mathfrak{h}_j$. Then $|D(h)| = |\pi_j(h)|^2$, $\forall$ $h \in \mathfrak{h}_j$. Thus there are constants C_j, (j = 1, 2, …, n) and normalized measures on $\mathfrak{g}$, and $\mathfrak{h}_j$, such that

$$f(g) = \int_{\mathfrak{g}} f(X)dX = \sum_j C_j \int_{\mathfrak{h}_j} |D(h)| \left(\int_{G/H_j} f(Ad(g)h)d(gH_j)dh_j \right)$$
$$= \sum C_j \int_{\mathfrak{h}'_j} |\Phi_f(h)| dh_j,$$

But $\mathfrak{g}_e \cong \mathbb{R}^n$, then

$$f(0) = \sum_j C_j \int_{\mathfrak{h}_j} |D(h)| \left(\int_{G/H_j} |D(Ad(g)h)|^{-1/2} f(Ad(g)h)d(gH_j)dh_j \right),$$

But $|D(Ad(g)h)|^{-1/2} = F_{p,q}(h)\varpi^{[n/2]} = F(h)\varpi^{[n/2]}$, then

$$\int_{G/H_j} |D(Ad(g)h)|^{-1/2} f(Ad(g)h)dgH = F(h)\varpi^{[n/2]} \int_{G/H_j} f(Ad(g)h)dgH,$$

Of fact, $|D|^{-1/2}$, is locally integrable on $\mathfrak{g}$, in the general case.

Then (I) can be written as:

$$f(0) = \sum_j C_j \int_{\mathfrak{h}_j} |\pi_j(h)|^2 \int_{G/H_j} f(Ad(g)h)dgHdh,$$

i

Then considering the cuspidal form of $\varpi^{[n/2]}f$, relative to the Cartan subgroup H_j, we have

$$\Phi^{H_j}_{\varpi^{[n/2]}f}(h) = E_j \varpi^{[n/2]} \int_{G/H_j} f(Ad(g)h)dgH,$$

The incise (i) take the form:

$$f(0) = \sum_j C_j \int_{\mathfrak{h}_j} |\pi_j(h)| \varepsilon_j(h) F(h) \Phi^{H_j}_{\varpi^{[n/2]}f}(h)dh,$$

But $\varpi_j^{[n/2]}\Phi_f^{H_j}(h) = \Phi_{\varpi}[n/2]_f^{H_j}(h)$, $\forall \varpi_j \in S(\mathfrak{h}_j)$. Then by the identity in appendix in [Wallach],

$$f(0) = \sum_j C_j \int_{\mathfrak{h}_j} |\pi_j(h)| \varepsilon_j(h) F(h) \varpi_j^{[n/2]} \Phi^{H_j}{}_f(h)dh,$$

Let $D_j = D_r(X)$,be with $r = \dim \mathfrak{h}_j$, $\forall\ X \in \mathfrak{g}$. Let $r = rangeX$, (assuming that $r \geq 2$)and let $n = \dim \mathfrak{g}$. To every $t > 0$, we define the space

$$\Omega_t = \{X \in \mathfrak{g} \mid |D_j(X)| < t,\ r \leq j < n\}, \qquad \text{(IX. 2.9)}$$

we consider the transcendent number $\pi = 3.14\ldots$Then we consider the lemma:

Lemma IX. 2.1 Suppose that G, is semisimple. If $0 < t < \pi - 1$, then $\exp|_{\Omega t}$, is a diffeomorphism.

Proof. Since to every $X \in \mathfrak{g}_C$, there is a neighborhood Ω_t, defined as the space

$$\Omega_t = \{X \in End(\mathfrak{g}_C) \mid |D_j(X)| < t,\ t > 0\}, \qquad \text{(IX. 2.10)}$$

If $j < \pi - 1$, then the map

$$\mathfrak{g}_C \rightarrow \exp(\mathfrak{g}), \qquad \text{(IX. 2.11)}$$

is a diffeomorphism of Ω_t, in a open subset of $I = [\mathfrak{g}, \mathfrak{g}]$, in $int(\mathfrak{g})$. Then given that dAd_e, is the homomorphic map

$$dAd_e : \mathfrak{g} \rightarrow \mathfrak{gl}(\mathfrak{g}), \qquad \text{(IX. 2.12)}$$

whose rule of explicit correspondence is

$$d(Ad_e(X)) = ad\ X, \quad \forall \quad X \in \mathfrak{g}, \qquad \text{(IX. 2.13)}$$

the map exp, restricted to Ω_t, result be a covering homomorphism that maps open sets in open sets of the space $\mathfrak{g}_C$, in $End(\mathfrak{g}_C)$, where these endomorphisms are differential operators. Therefore $\exp|_{\Omega t}$, is a diffeomorphism. ■

Let W, be an open neighborhood of O, in $z(\mathfrak{g})$, such that $\exp|_W$, is a diffeomorphism. Let Ω_t, be an open set in the ideal $[\mathfrak{g}, \mathfrak{g}]$. Let $W_t = W \oplus \Omega_t$, be. Then

1. If $0 < t < \pi - 1$, then exp, is a diffeomorphism of W_t, in an open neighborhood V_t, of 1, in G.

Let $u \in C_c^{\infty}(\mathbb{R})$, $0 \leq u(s) \leq 1$, be such that $u(s) = 1$, to $s \leq (\pi - 1)/2$, and $u(s) = 0$, to $s > 2(\pi - 1)/3$. Let $O \in Y \subset Cl(Y)$ and $Cl(Y) \subset W$, with open Y, and Cl(Y), compact. Let $h \in C_c^{\infty}(W)$, be with $h(X) = 1$, to $X \in Cl(Y)$. Then we define the function $\beta \in C^{\infty}(z)$, such that

$$\beta : W_{\pi-1} \rightarrow C^{\infty}(W), \qquad \text{(IX. 2.14)}$$

with rule of correspondence

$$X = Z + T \mapsto \beta(X), \tag{IX. 2.15}$$

where explicitly

$$\beta(X) = h(Z)\prod_{r \leq j \leq n-1} u(D_j(T)), \tag{IX. 2.16}$$

Outside $W_{\pi-1}$, $\beta = 0$. Then

2. $\beta \in C^{\infty}(\mathfrak{g})$, $\text{supp}\, \beta \subset W_{\pi-1}$.

β, is Ad(G)-invariant, since

3. $\beta(Ad(\mathfrak{g})X) = \beta(X)$, $\forall$ $X \in \mathfrak{g}$, and $g \in G$.
4. If $\mathfrak{h}$, is a Cartan subalgebra of $\mathfrak{g}$, then $\text{supp}\, \beta \cap \mathfrak{h}$, is compact.

Indeed, the partitions of the unity u and h, have supports included in W, and $\Omega_{\pi-1}$, respectively. Given that $W_{\pi-1} = W \oplus \Omega_{\pi-1}$, then $\text{supp}\, \beta \subset W \cap \Omega_{\pi-1}$. Then $h\prod u D_j \in C^{\infty}(\mathfrak{g})$.

The Ad(G)-invariance is followed of the existence of the diffeomorphism $\exp|_{W_{\pi-1}}$, which is a covering homomorphism of int(G). The last affirmation is followed [Chevalley]. We will introduce a function α, on G, defined as the map

$$\alpha : G \to C^{\infty}(W_{\pi-1}), \tag{IX. 2.17}$$

with rule of correspondence

$$\mathfrak{g} \mapsto \alpha(\mathfrak{g}), \tag{IX. 2.18}$$

and such that applying the diffeomorphism $\exp|_{W_{\pi-1}}$, to the image $\alpha(\mathfrak{g})$, is had that:

$$\alpha(\exp X) = \beta(X), \ \forall \ X \in \mathfrak{g},$$

inside $W_{\pi-1}$, and $\alpha = 0$, outside $W_{\pi-1}$.

Such map is smooth on G, since $\alpha(g) \in C^{\infty}(G)$, and $\alpha(gxg^{-1}) = \alpha(x)$, $\forall$ x, $y \in G$. Then, if $f \in C^{\infty}(G)$, then $C^{\infty}(W)f = f^{\sim}$, where $f^{\sim}(X) = \beta(X)f(\exp X)$, $\forall \beta(X) \in C^{\infty}(W)$, and $X \in \mathfrak{g}$. Clearly $f^{\sim}(0) = \beta(0)f(1)$. But $\beta(0) = 1$, $\forall$ $s < \pi - 1$, since $u(s) = 1$, $\forall$ $s \leq (\pi - 1)/2$, and $h(s) = 1$ in all $Cl(W_{\pi-1})$. For other side $f^{\sim}(0) = f(1)$.

Consider H, a Cartan subgroup of G. Then we can write to $\Delta_H(h)\Phi^H_{f\sim}(h)$, as (using $\beta(h)$)

$$\Phi^H_{\tilde{f}}(H) = \frac{\varepsilon_H(h)}{\Delta_H(h)}\beta(h)F^H_f(\exp h), \tag{IX. 2.19}$$

$\forall$ $f \in C_c^{\infty}(G)$, and $h \in \mathfrak{h}''$.

Since $\Delta_H(\exp h)/\pi(h) \neq 0$, $\forall\ h \in W_{\pi-1} \cap \mathfrak{h}$, then $\pi(h)/\Delta_H(\exp h)$, define a smooth family on $W_{\pi-1} \cap \mathfrak{h}$, and since the map given by $f \mapsto F_f^H$, is extended to a continuous map of $\mathscr{C}(G)$ in $\mathscr{C}(H'')$, we have that

5. The map $f \mapsto \Phi_{\tilde{f}}^H$, is extended to a continuous map of $\mathscr{C}(G)$ in $C_c^\infty(\mathfrak{h}'')$.

Then

$$f(0) = \sum_j c_j \int_{\mathfrak{h}_j'} \left|\pi_j(h)\right| \varepsilon_j(h) F(h) \varpi_j^{[n/2]} \Phi_f^{H_j}(h) dh, \qquad \text{(IX. 2.20)}$$

implies that if $f \in \mathscr{C}(G)$, then

$$\begin{aligned} f(1) &= \sum_j c_j \int_{\mathfrak{h}_j'} \pi_j(h) F(h) \varpi_j^{[n/2]} \Phi_{\tilde{f}}^{H_j}(h) dh, \\ &= \sum_j c_j \int_{\mathfrak{h}_j'} \pi_j(h) F(h) \varpi_j^{[n/2]} \frac{\pi_j(h)}{\Delta H_j(\exp h)} \beta(h) F_f^H(\exp h) dh, \end{aligned} \qquad \text{(IX. 2.21)}$$

Likewise (IX. 2. 21) implies that if $f \in {}^0\mathscr{C}(G)$, and if G, include non-compact subgroups (where $G = {}^0G$) then $f(1) = 0$. Now if $f \in {}^0\mathscr{C}(G)$, then $R(g)f \in {}^0\mathscr{C}(G)$. Therefore, if G include non-compact subgroups then to ${}^0\mathscr{C}(G)$, is necessary the null space {0}. This last demonstrate the first part of the theorem. ■

Now we demonstrate that: If T is a Cartan compact subgroup of G then there is a constant $c_G \neq 0$, such that $\forall\ f \in {}^0\mathscr{C}(G)$,

$$\varpi F_f^T(1) = c_G f(1),$$

$\forall\ F_f \in C^\infty(T)$.

We assume that $H_j = T$, and $f \in {}^0\mathscr{C}(G)$, then (IX. 2. 21) takes the form

$$f(1) = c_j \int_{\mathfrak{t}} \pi_j(h) F(h) \varpi^{[n/2]} \frac{\pi(h)}{\Delta_t(\exp(h))} \alpha(h) F_f^T(\exp h) dh, \qquad \text{(IX. 2.22)}$$

$\forall\ F_f \in C^\infty(T)$.

We consider the following lemma that will be essential in the demonstration of the second affirmation of the theorem IX. 2. 1.

Lemma IX. 2.2. There is a constant $M_g \neq 0$, such that if $g \in A(\mathfrak{t})$, then

$$\int_{\mathfrak{t}} \pi_j(h) F(h) \varpi^{[n/2]} g(h) dh = M_g g(0), \qquad \text{(IX. 2.23)}$$

Note: M_g, will be calculated in the course of the demonstration.

Let $p \in P(t_C)$, be such that $p = D^0(X)$, $\forall$ $X \in \mathbf{t}$. Let X, $Y \in End(V)$, with V, a vector space. Then are valid the following commutation identities:

i $$[X^k, Y] = \sum_{j=0}^{k-1} \binom{k}{j} (-1)^{k-j+1} X^j((adX)^{k-j}Y) = \sum_{j=0}^{k-1} \binom{k}{j} ((adX)^j Y)X^{k-j}, \qquad (IX.\ 2.24)$$

If we consider $[\varpi^{[n/2]}, \pi] = \varpi^{[n/2]}\pi - \pi\varpi^{[n/2]}$, then sustituying (IX. 2. 23), we have

$$\int_t F(h)\pi(h)\varpi^{[n/2]}g(h)dh = \int_t F(h)\varpi^{[n/2]}\pi(h)g(h)dh - \int_t F(h)[\varpi^{[n/2]}, \pi]g(h)dh$$
$$= I - II,$$

Then if T, is a Cartan subgroup then p = dim$\mathfrak{p}$, and q = dimt, are pairs (q = dim$\mathfrak{T}$). By the theorem 7.A 5. 8, there is a not null constant$B_g \neq 0$, such that

$$I = B_g(\varpi^{[n/2] - [r, 2]})(\pi g)(0), \qquad (IX.\ 2.25)$$

Then is necessary calculate the integral II. Note that $n - r$, is even then $[n/2] - [r/2] = (n - r)/2$. Therefore by i):

$$[\varpi^{[n/2]}, \pi] = \sum_{j=0}^{[n/2]-1} \binom{[n/2]}{j} (-1)^{[n/2]-j+1} \int_t F(h)\varpi^j((ad\varpi^{[n/2]})^j \pi)g(h)dh, \qquad (IX.\ 2.26)$$

Therefore $(ad\varpi)^{[n/2] - j}\pi = 0$, if $j < [r/2]$. Then newly we apply the theorem 7. A. 5. 8, and we find that II, is expressed as:

$$II = \int_t F(h)[\varpi^{[n/2]}, \pi]g(h)dh = \sum_{j=0}^{[n/2]-1} \binom{[n/2]}{j}$$
$$(-1)^{[n/2]-j+1} B_g((\varpi^{j-[r/2]}(ad\varpi)^{[n/2]-j}\pi))g(0), \qquad (IX.\ 2.27)$$

Appliying the second identity in (IX. 2. 21) in terms of the adjunct map "ad" and observing that the coefficients of ad, are annulled to $j > [r/2]$ (Escolium 7. A. 2. 9) is had to $j \geq [r/2]$ that

$$\varpi^{j - [n/2]}(ad\varpi^{[n/2] - j}\pi)g(0) = (ad\varpi^{[n/2] - [r/2]}\pi)g(0),$$

Then

$$\int_t F(h)\pi(h)\varpi^{[n/2]}g(h)dh = B_g(\varpi^{[n/2]-[r/2]})(\pi g)(0) - B_g(ad\varpi^{[n/2]-[r/2]}\pi)g(0)$$
$$= C((ad\varpi^{[n/2]-[r/2]}\pi)g)(0),$$

where

$$C = B_g\left(\sum_{j=[r/2]}^{[n/2]}\binom{[n/2]}{j}(-1)^{[n/2]-j}\right) \quad \text{(IX. 2.28)}$$

But the scolium 7. A. 2. 9, implies that

$$\mathrm{ad}\varpi^{[n/2]-[r/2]}\pi = 2^{[n/2]-[r/2]}([n/2]-[r/2])\Pi_{\alpha\in\Phi^+}H_\alpha,$$

From the second idendity in (IX. 2. 21) and for the identity

$$\sum_{j=p}^{k}(-1)^{k-j}\binom{k}{j} = (-1)^{k-p}\binom{k-1}{j-1} \neq 0, \quad \text{(IX. 2.29)}$$

if $k \geq p > 0$, we have that

$$\int_t F(h)\pi(h)\varpi^{[n/2]}g(h)dh = \left[\sum_{j=p}^{[n/2]}(-1)^{k-p}\binom{k-1}{p-1}2^{[n/2]-[r/2]}([n/2]-[r/2])!\prod_{\alpha\in\Phi^+}H_j\right]g(0)$$
$$= M_g g(0),$$

■

Also we use the estimation of $\int_t F(h)\pi(h)\varpi^{[n/2]}g(h)dh$, through of a smooth function u, that is W-invariant defined on a W-invariant neighborhood of the 0.

Lemma IX. 2.3. Let $W = W(\mathfrak{g}_C, t_C)$. Let $u\in C^\infty{}_W(U_e{}^W)$. Then

$$\sum_{F\in\Phi^+}\left(\left(\prod_{\alpha\notin F}H_\alpha\right)u\right)(0)\prod_{\alpha\in F}H_\alpha = u(0)\prod_{\alpha\in F}H_\alpha, \quad \text{(IX. 2.30)}$$

Proof. Let α, be a simple root in Φ^+. If $F \subset \Phi^+$, then we define the map ℘:

$$\wp : F \to F^{\wp}, \quad \text{(IX. 2.31)}$$

with rule of correspondence

$$\alpha \mapsto \alpha^{\wp}, \quad \text{(IX. 2.32)}$$

and where

$$F^{\wp} = \{s_\alpha F, \text{ if } \alpha\notin F\ (\alpha\in F^{\wp}\setminus F) \text{ and } (s_\alpha(F-\{\alpha\}) \cup \{\alpha\}, \text{ if } \alpha\in F, \quad \text{(IX. 2.33)}$$

Then the map ℘, is a bijection of $P(\Phi^+)$. Indeed, if we consider that the dense space or kernel is the set

$$N(\wp) = \{\alpha\in F \mid \wp\alpha = \alpha^{\wp} = 0\}, \quad \text{(IX. 2.34)}$$

and since α, is simple root then $\alpha \neq 0$. Then $\ker \wp \neq \{0\}$. Therefore $\wp$, is injective. Also is suprajective, since, let

$$p = \int_{t} F(h)\pi(h)\varpi^{[n/2]}g(h)dh\ ,$$

Then $s_\alpha p$, comes given as:

$$s_\alpha p = p^{\wp} = \left[\sum_{F\in\Phi^+}\left(\left(\prod_{\alpha\notin F} H_\alpha\right)u\right)(0)\prod_{\alpha\in F} H_\alpha\right]^{\wp} \qquad \text{(IX. 2.35)}$$

Given that $sp = p^{\wp} = -p$, $\forall p\in P(t_C)$, then (IX. 2. 34) takes the form

$$p^{\wp} = -\sum_{F\in\Phi^+}\left(\left(\prod_{\beta\notin F^{\wp}} H_\beta\right)u\right)(0)\prod_{\beta\in F^{\wp}} H_\beta = -p, \qquad \text{(IX. 2.36)}$$

Where is had used strongly the property $xu(0) = (sx)u(0)$. $\forall\ x\in S(\mathbf{t}_C)$. Then $sp = \det(s)p$, $\forall$ $s\in W$, which implies that $p = q\prod_{\alpha\in\Phi^+} H_\alpha$, $\forall q\in S(\mathbf{t}_C)$. Given that

$$p = u(0)\prod_{\alpha\in\Phi^+} H_\alpha = q\prod_{\alpha\in\Phi^+} H_\alpha, \quad \forall\ \ q\in S(\mathbf{t}_C) \qquad \text{(IX. 2.37)}$$

then $q = u(0)$. Therefore

$$\sum_{F\subset\Phi^+}\left(\left(\prod_{\alpha\notin F} H_\alpha\right)u\right)(0)\prod_{\alpha\in F} H_\alpha = u(0)\prod_{\alpha\in\Phi^+} H_\alpha .$$

Applying IX. 2.1 (1) and the lemma IX. 2. 2, we have

$$f(1) = c_j M_g\left(\prod_{\alpha\in\Phi^+} H_\alpha\right)\frac{\pi(h)}{\Delta(\exp(th))\big|_{t=1}}\beta(h)F_f^T(\exp(th))\Big|_{h=0,t=1,\beta=1}, \qquad \text{(IX. 2.38)}$$

Considering $u(h) = \pi(h)/(\Delta(\exp(h))$, and given that $\forall\beta\in C^\infty(\mathfrak{g})$, with $u = 1$, in $U_e = W_{\pi-1}$. Then $u\in C^\infty(W_{\pi-1})^W$, and $u(0) = 1$. Then

$$f(1) = c_j M_g \varpi u(0)\beta(0)F_f^T(1),$$

or equivalently $\varpi F_f^T(1) = C_G f(1)$. ■

IX. 3. Integral transforms in algebraic analysis

In this section, we state some of the foundational results we develop in order to study Hecke categories. A key idea of algebraic analysis is to replace the functions and distributions of harmonic analysis by the algebraic systems of differential equations that they satisfy. One can view this as a form of categorification where the resulting *D*-modules play the role of the original functions and distributions, and categories of *D*-modules play the role of generalized function spaces. For example, the exponential function$f(x) = e^{\lambda x}$, is characterized by the algebraic equation $(\partial x - \lambda)f(x) = 0$,and hence is a solution of the *D*-module $\mathcal{D}_{\mathbb{A}}1/\mathcal{D}_{\mathbb{A}}1(\partial x - \lambda)$. Similarly, the delta distribution $\delta\lambda$, is

characterized by the algebraic equation $(x - \lambda)f(x) = 0$, and hence is a solution of the D-module$D_{\mathbb{A}}1/D_{\mathbb{A}}1(x - \lambda)$.Natural operations in harmonic analysis are given by integral transforms acting on function spaces

$$f(x) \mapsto (K * f)(y) = \int f(x)K(x, y)dx, \qquad \text{(IX. 3.1)}$$

where $K(x, y)$, is an integral kernel. For example, (one normalization of) the Fourier transform on the real line is given by the integral kernel $K(x, y) = e^{-2\pi ixy}$. In continued analogy, natural operations in algebraic analysis are given by integral transforms acting as functors between categories of *D*-modules.In this context, derived versions of tensor product and push forward replace multiplication and integration respectively. To be more precise, given varieties X, Y, and a *D*-module $\mathcal{K}$, on the product X × Y, one defines a functor on derived categories of *D*-modules

$$\mathcal{D}(X) \rightarrow \mathcal{D}(Y), \qquad \text{(IX. 3.2)}$$

with the rule of correspondence

$$\mathcal{F} \mapsto \pi_{Y*}(\pi^{\wedge}{}_{X}\mathcal{F} \otimes \mathcal{K}), \qquad \text{(IX. 3.3)}$$

by pulling back from X, to the product X × Y, tensoring with the integral kernel $\mathcal{K}$, and then pushing forward to Y, via the natural diagram

$$X \xleftarrow{\pi_X} X \times Y \xrightarrow{\pi_Y} Y, \qquad \text{(IX. 3.4)}$$

For example, the geometric Fourier transform of Malgrange, an autoequivalence of D-modules on $\mathbb{A}^1$, is given by the integral kernel $\mathcal{K} = \mathcal{D}_{\mathbb{A}^{x1} \times_{\mathbb{A}^{y1}}} / \mathcal{D}_{\mathbb{A}^{x1} \times_{\mathbb{A}^{y1}}}(\partial_x - iy)$, with solution $K(x, y) = e^{ixy}$. The classical Fourier transform of a solution of a *D*-module $\mathcal{F}$, is a solution of the geometric Fourier transform of $\mathcal{F}$.

For another example, given a correspondence of varieties

$$X \xleftarrow{f} Z \xrightarrow{g} Y, \qquad \text{(IX. 3.5)}$$

one defines a functor on derived categories of *D*-modules by the similar formula from

$$\mathcal{D}(X) \rightarrow \mathcal{D}(Y):$$

$$\mathcal{F} \mapsto g_* f^{\wedge} \mathcal{F}), \qquad \text{(IX. 3.6)}$$

By the projection formula, this functor coincides with the integral transform given by the integral kernel $\mathcal{K}= (f \times g)_* \mathcal{O}_Z$ on the product X × Y. In general, integral transforms can be interpreted as operations on systems of differential equations, transforming solutions to one system into solutions of a new (and potentially more accessible)system. The theory of integral transforms for *D*-modules has been developed and applied to a host of problems in integral geometry and analysis, in particular to the study of the Radon, Laplace and Penrose transforms, starting with the influential paper of Brylinski [Br] and continuing in the beautiful work of Goncharov, Kashiwara, Schapira, D'Agnolo [22, 41] and others. The *D*-module approach allows one to separate the algebraic and geometric aspects underlying a system of differential equations from the analytic problems involving solvability in different function spaces, allowing one to obtain powerful general results.

IX. 4. Revisited integrable square representations

Consider to G, a real reductive group of inner type and such that ${}^0G = G$. Consider the space $E_2(G)$, of equivalence classes of irreducible square integrable representations of G, that is to say, the space

$$E_2(G) = \{\sigma \in [\sigma] \mid |\pi_\sigma(g)|^2 < \infty, \forall g \in G, \text{ and } (\pi_\sigma, H_\sigma) \in \sigma\}, \qquad \text{(IX. 4.1)}$$

Where

$$\|\pi_\sigma(g)\|^2 = \int_G |\pi_\sigma(g)|^2 dg < \infty, \ \forall \ g \in G,$$

If v, $w \in (H_\sigma)_K$, then the coefficients to the classes $[\sigma]$, comes given by the map

$$c_{v,w}: G \rightarrow \mathscr{C}(G), \qquad \text{(IX. 4.2)}$$

whose rule of correspondence is:

$$g \mapsto < \pi_\sigma(g)v, w >, \qquad \forall \quad v, w \in (H_\sigma)_K \qquad \text{(IX. 4.3)}$$

where $c_{v,w}(g) \in \mathscr{C}(G)$, since $(H_\sigma)_K$, is rapidly increasing which is also a finite $Z(\mathfrak{g})$- module ($\dim Z_G(\mathfrak{g}) c_{v,w} < \infty$). Then $c_{v,w}$, is a cuspidal form on G. Therefore $c_{v,w}(g) \in {}^0\mathscr{C}(G)$. Then is had the theorem:

Theorem IX. 4.1. $E_2(G)$, is non-vanishing if and only if G, have a compact subgroup.

Proof. First we will demonstrate the implication ⇒). If $E_2(G) \neq \emptyset$, then exist a integrable square representation $(\pi_\sigma, H_\sigma) \in \sigma$, such that

$$\|c_{v,w}(g)\|^2 = \int_G|<\pi_\sigma(g)v, w>|^2dg < \infty, \quad \forall \; C_{v,w} \in {}^0\mathscr{C}(G),$$

Then ${}^0\mathscr{C}(G) \neq \{0\}$. Therefore G, not include non-compact subgroups of Cartan. Then G, have only compact subgroups of Cartan.

For other side, if G, have Cartan subgroups of G, then by the fundamental theorem of Harish-Chandra[8], $E_2(G) \neq \varnothing$. Then we assume that T, is one of such Cartan compact subgroups of G. We write to T, as $T = ZT^0$.

Let the character ζ_μ, as the map or homomorphism

$$\zeta_\mu : Z(\mathfrak{g}) \rightarrow \mathbb{C}, \tag{IX. 4.4}$$

with rule of correspondence

$$z \mapsto \zeta_\mu(z) = \mu \circ \gamma(z), \tag{IX. 4.5}$$

where $\gamma(t)$, is the Harish-Chandra isomorphism associated with T. Likewise, if we consider $\mu|_{T0}$, and given that $\mu|_{T0} = d(\mu)\Lambda(\mu)$, and $\zeta_\mu = d(\mu)\mu$, then

$$\zeta_\mu(z) = \Lambda(\mu) \circ \gamma(z) = \Lambda\mu \circ F^T{}_{zf}(\mu), \tag{IX. 4.6}$$

If $f \in {}^0\mathscr{C}(G)$, then $F_f \in C^\infty(T)$. Endeed, given that $F_f{}^T(t)$, is the continuous map:

$$F_f{}^T : {}^0\mathscr{C}(G) \rightarrow C^\infty(T), \tag{IX. 4.7}$$

with rule of correspondence

$$f \mapsto F_f{}^T(t) = \Delta_T(t)\int_{T'} f(gtg^{-1})dt, \tag{IX. 4.8}$$

with T′, an orbit of T, and since $F^T{}_{zf}(t) = \gamma(t) \circ F_f{}^T(t)$, then $F_f{}^T(t) \in C^\infty(T)$. Then the Peter-Weyl theorem implies that $\forall \mu \in T^\wedge$,

i. $F_f = \Sigma_{\mu \in T}(F_f)^\wedge(\mu)\zeta_\mu$,

If in particular we consider a $h \in C^\infty(T)$, T-central then

$$h^\wedge(\mu) = \int_T h(t)\mathrm{conj}(\zeta_\mu(t))dt,$$

Then the second part of the theorem IX. 2. 1, implies that exist $C_G \neq 0$, (constant) such that

$$\varpi F_f{}^T(1) = C_G f(1),$$

or equivalently

[8] **Theorem.** If G, contains a compact Cartan subgroup then G, has irreducible square integrable representations

i. $f(1) = \tilde{C}_G(\varpi F_f^T)(1) = \tilde{C}_G \sum_{\mu \in T^\wedge} \left(\prod_{\alpha \in \Phi^+} (\Lambda(\mu), \alpha) \right) (F_f^\wedge)(\mu),$

Then, if $z \in Z(\mathfrak{g}_C)$, then we have that

ii. $F_{zf} = \gamma(z) F_f,$

on T′, and therefore on T, or their corresponding dual:

$$(F_{zf})^\wedge(\mu) = \Lambda(\mu)(\gamma(z))(F_f^\wedge(\mu)) = \Lambda(\mu)(F_{zf}^\wedge(\mu)),$$

Which is correct, since $\Lambda(\mu)$, is a character $T^0 = \{t \in T \mid Ad(t)F_f^T = F_f^T, \ \forall \ F_f^T \in C^\infty(T)\}$. Therefore finally and considering the map

$$\chi_\sigma \colon Z(\mathfrak{g}_C) \to \mathbb{C}, \qquad \text{(IX. 4.9)}$$

that extends as infinitesimal character of σ, $\forall \sigma \in E_2(G)$, to ζ_μ, from T′, to T, then

$$Z(\mathfrak{g}_C) f = f, \quad \forall \ f \in {}^0\mathscr{C}(G), \qquad \text{(IX. 4.10)}$$

of where

$$zf = \chi_\sigma(z) f, \qquad \text{(IX. 4.11)}$$

Thus has been demonstrated that:

Theorem IX. 4.2. Let $\sigma \in E_2(G)$, be, then there is $\mu \in T^\wedge$, such that $(\Lambda(\mu), \alpha) \neq 0$, $\forall \alpha \in \Phi(\mathfrak{g}_C, \mathfrak{t}_C)$, and such that the infinitesimal character of σ, is $X_{\Lambda(\mu)}$.

From the before theorem is deduced the following corollary:

Corollary IX. 4.1. Let $\gamma \in K^\wedge$, be then number of classes $\sigma \in E_2(G)$, such that $(H_\sigma)_K(\gamma) \neq 0$, is finite.

Proof. Consider the Casimir operator corresponding to B. Let $C_K = C|_{\mathfrak{k}}$, to K, corresponding to B. Let $X_1, \ldots, X_n$, be an orthogonal base of $\mathfrak{p}$, realative to B. Let

$$C_{\mathfrak{p}} = \Sigma (X_i)^2, \qquad \text{(IX. 4.12)}$$

Then $C = C_K + C_{\mathfrak{p}}$. Fixing $\gamma \in K^\wedge$, and considering μ_γ, an eigenvalue of C_K, on any representant of the class γ, we note that

a. If (π, H), is an unitary representation of G, with C, acting for cI, and if $H_K(\gamma) \neq 0$, then $c \leq \mu_\gamma$.

Endeed, if $v \in H_K$, such that $\|v\| = 1$, then

$$c = c\langle v, v\rangle = \langle Cv, v\rangle = \langle C_K v, v \rangle = \langle C_{\mathfrak{p}} v, v\rangle = \mu_\gamma - \Sigma \langle X_j v, X_j v\rangle \leq \mu_\gamma,$$

If $\sigma \in E_2(G)$, then Λ_σ, is the character of $(T^0)^\wedge$, such that $\Lambda_\sigma - \mu_\gamma = \chi_\sigma(C)$.

Let $\rho = \frac{1}{2}\sum_{\alpha\in\Phi^+}\alpha$, be, then

$$\chi_\sigma(C) = \|\Lambda_\sigma\| - \|\rho\|, \tag{IX. 4.13}$$

For

$$c = \mu_\gamma - \Sigma\langle X_j v, X_j v\rangle, \ \forall \ v \in H_K(\gamma), \tag{IX. 4.14}$$

such that $\|v\| = 1$. Thus

$$\|\Lambda_\sigma\|^2 \leq \|\rho\|^2 + \mu_\gamma, \tag{IX. 4.15}$$

Then the unique number of possibilities to obtain an infinitesimal character of integrable square whose γ–isotopic component $H_K(\gamma) \neq 0$, is $c \leq \mu_\gamma$, that is to say, whose infinitesimal character $\chi_\sigma(C)$, let be such that $\Lambda_\sigma - \mu_\gamma = \chi_\sigma(C)$.

Since there is an equivalence between the finite number of irreducible $(\mathfrak{g}, K)$-module with infinitesimal character χ_σ, in the class σ, then the number of $\sigma \in E_2(G)$, such that $H_K(\gamma) \neq 0$, is finite. ■

Chapter X

Cohomological Induction and Securing of Generalized G-Modules by G(w)-Orbits to Infinite Dimensional Representations of Lie Groups

(Talk given in Ivano-Frankivsk, Ukraine, 2009 [15])

X. 1. Vogan program

In the study by Vogan, is to establish the following problem of the representation theory (PTR). Given an irreducible representation (π, V), it is possible to give a structure of Hilbert space on V, to have to π, like an unitary representation.

Strictly speaking this bears the questions:

i. We can make that V, takes a bilinear hermitian G- invariant form $< , >_{\pi}$?
ii. If whose form exist, ¿Is the form $< , >_{\pi}$, positive define?

The goal of the Vogan program is to analyze some difficulties that arise when one tries to study this program accurately this program. The difficulties have their origin exactly in the flexibility of the definition of a representation (π, V), of G, a topological group.

Typically wants to want realize a representation of G, on a space of functions. If G, acts on the set X,then G, it acts on functions on X, for

$$\left[\pi(g)f\right](x) = f(g^{-1}x), \ \forall x \in X \text{ and } g \in G, \qquad (X.\ 1.1)$$

The difficulties arise when we try to decide exactly which space of functions on X, it is necessary to consider. Since if G, is the Lie group acting smoothly on a manifold X, then one can consider $C(X)$, $C_c(X)$, $C_c^{\infty}(X)$ or $C^{-\omega}(X)$.

Harish - Chandra establish that: Each class of infinitesimal equivalence of admissible irreducible representations contains at most a class of equivalence of irreducible unitary representations of G, this is:

$$\tilde{G}_u \subset \tilde{G}, \qquad (X.\ 1.2)$$

INTECH
open science | open minds

with $\tilde{G}$, a class of infinitesimal equivalence of classes of admissible representations of G. This result in some way establishes a solution or general answer to the problem (PTR) of the Vogan program.

However, requires to specific the type in a positive defined way that will guarantee the Hermitian structure of the representation algebra of V_K, that induces to $< , >_{\pi, K}$, like a Hermitian G-invariant form and thus endows to V_K, like unitary representation in this class of infinitesimal equivalence, that is to say, the existence will have been guaranteed of at least an irreducible unitary representation in the class of unitary equivalence that is representation of G.

X. 2. Dificulties

The existence in a G-invariant continuous Hermitian form $< , >_{\pi}$, on V, implies the existence of $< , >_{\pi, K}$, on V_K, but the reciprocal is not true. Since V_K is dense in V, exists a continuous extension at most of $< , >_{\pi, K}$, to V, but the extension cannot exist.

An important observation that one deduces from some concrete examples is that the Hermitian form can be defined only on "representations appropriately thin or small", at least it is what Vogan puts in evidence inside its program for each class of infinitesimal equivalence.

In the different spaces to consider $C(X)$, $C_c(X)$, $C_c^{\infty}(X)$ ó $C^{-\omega}(X)$, the space $C_c^{\infty}(X)$, offers as a suitable candidate and appropriately small and it can be endowed with a invariant Hermitian form. The space is generally more "fat" like space to admit a invariant Hermitian form.

We define the space V*, of continuous linear functionals on V, endowed of the strong topology, that is

$$V^* = \{\xi \in C(V) | \xi : V \rightarrow C, \tau(\xi) : W_{\varepsilon}(B)\}$$

$$= \Im(V, C), \tag{X. 2.1}$$

$\tau(\xi)$, is the defined strong topology on neighborhood base in the origin consisting of groups $W_{\varepsilon}(B)$, defined by the spaces

$$W_{\varepsilon}(B) = \{\xi \in E^* | \sup_{e \in B} |\xi(e)| \leq \varepsilon\} \subset E^*,$$

with $B \subset V$, bounded.

Theorem X. 2.1. (Casselman, Wallach y Schmid) [46-48].

Suppose that (π, V), is an irreducible admissible representation of Lie groups G, on a Banach space V. We Define

$$(\pi^{\omega}, V^{\omega}) = analitic.vectors.in.V,$$
$$(\pi^{\infty}, V^{\infty}) = differentiable.vectors.in.V,$$
$$(\pi^{-\infty}, V^{-\infty}) = distribution.vector.in.V$$
$$= dual.de.(V')^{\infty}$$
$$(\pi^{-\omega}, V^{-\omega}) = vectors.of.hyperfunctions.in.V$$
$$= dual.of.(V')^{\omega}.$$

Each one of these four representations is a soft representation of G, in the class of infinitesimal equivalence of π, and each one only depends on that equivalence class.

The inclusions

$$V^{\omega} \subset V^{\infty} \subset V \subset V^{-\infty} \subset V^{-\omega}, \qquad (X.\ 2.2)$$

are continuous, with dense image. Anyone Hermitian form $< , >_K$, on V_K, expands uniquely to G-invariants continuous Hermitian forms $< , >_{\omega}$, and $< , >_{\infty}$, on V^{ω}, and V^{∞}.

The four representations $V^{\omega}, V^{\infty}, V^{-\infty}$ and $V^{-\omega}$, are called minimal, smooth, distribution and maximal globalizations respectively.

Except for π, be ν, a representation of finite dimension (such that all the spaces in the theorem X. 1, they are the same one). The Hermitian form cannot extend continually to the maximal distribution or globalization $V^{-\infty}$, and $V^{-\omega}$. For what could be necessary the use of representations of G, built on spaces of holomorphic sections of vector bundles and generalizations. Is in this part where later on inside this work it will be to induce and to generalize the G-modules of Harish-Chandra [5, 21] and [37], to be able to be related with the globalizations of Wong.

Now then, since through this way are obtained unitary representations, is necessary to specify a similar way or analogous to the followed to the obtaining of minimal and differentiable globalizations with the certainty of that the Hermitian forms can be defined on the representations.

X. 3. Minimal and maximal globalizations

We consider a result of maximal globalizations of Harish-Chandra:

Theorem X. 3.1. (Wong) [36]. We assume that the admissible representation V, is the maximal globalization of the ($\mathfrak{q}$, L∩K)-module underlying. Be the G-invariant holomorphic vector bundle on X = G/L, corresponding of $(V \to X)$. Then the operator $\bar{\partial}$, for the Dolbeault cohomology has closed range and such that each one of the spaces $H^{p,q}(X,\nu)$ takes a soft representation of G. Each one of these admissible representations is the maximal globalization of their underlying module of Harish-Chandra.

Proof: [17].

Def. III.1. Suppose G, is real and reductive. $\mathfrak{q}$,is a parabolic subalgebra of the complexified Lie algebra $\mathfrak{g}_C$, and L, is the Levi factor of$\mathfrak{q}$. A ($\mathfrak{q}$, L)-representation (τ, V) is to say be admissible if the representation τ, in L, is admissible. In this case the method of Harish-Chandra of V, is the ($\mathfrak{q}$, L$\cap$K)-module $V^{L\cap K}$, of vectors L$\cap$K-finites in V.

The theorem of Wong [36], establish that the Dolbeault cohomology let to the maximal globalizations in great generality. This means that there is not possibility to find Hermitian invariant forms on these representations of Dolbeault cohomology except in the case of finite dimension. That is to say, spaces are obtained too "fat" to be able to identify and to classify the infinitesimal equivalence classes of representations of Lie groups and to identify the unique classes of unitary representations corresponding to each one of the mentioned infinitesimal equivalence classes.

For it is necessary to develop a way to modify the Dolbeault cohomology to produce minimal globalizations in more grade that the maximal. Essentially we can follow the ideas of Serre, which are based on the realization of representations of minimal globalization obtained about generalized flag manifolds achieved first by Bratten. Of the duality used to define the maximal globalization the question it does arise, How can you identify the dual topological space of a cohomological space of Dolbeault on a complex compact neighborhood?

The question is interesting in the simplest case: Suppose an $X \subset C$, is open set and H(X), is the space of holomorphic functions X, in a topological vector space X, using the topology of uniform convergence of all the derivatives on compact sets.

For what would H(X), be natural alone to be questioned, what is the dual space?

This last question has a simple answer. Be $C_c^{-\infty}(X, \textit{densidades})$, the space of supported distributions compactly on X. We can think in this like the space of complex supported 2-forms compactly (or (1, 1)-forms) on X, with coefficients of generalized functions. For what, with more generality we write:

$A_c^{(p,q),-\infty}(X) = (p,q) -$ Supported forms compactly on X, with coefficients of generalized functions,

The differential operator of Dolbeault $\overline{\partial}$, map (p, q)-forms to (p, q + 1)-forms and it preserves the support,

$$\overline{\partial} : A_c^{(1,0),-\infty}(X) \rightarrow A_c^{(1,1),-\infty}(X) = C_c^{-\infty}(X, \mathit{densities}) \tag{X. 3.1}$$

Then

$$H(X) \cong A_c^{(1,1),-\infty}(X) / \overline{\overline{\partial} A_c^{1,0}(X)}, \tag{X. 3.2}$$

Here the line ——on $\overline{\partial} A_c^{1,0}(X)$, it denotes closing of the space $\overline{\partial} A_c^{1,0}(X)$.

To open X, in C, the image of $\overline{\partial}$, is automatically closed, for that the line—— is not necessary. However this formulation has an immediate extension for any complex manifolds X (replacing 1, and 0, for the dimensions n, and $n-1$).

Let us enunciate the generalization of Serre [13]:

Theorem X. 3.2. (Serre). Suppose that X, a complex manifold of dimension n, ν a vector holomorphic bundle on X, and Ω, is the canonical bundle of lines (of (n, 0)-forms have more than enough X). We Define

$$A^{0,p}(X, \nu) = \mathit{Space\ of\ the\ Soft\ } \nu\mathit{-valuates\ (0,\ p)\text{-}forms\ on\ X,} \tag{X. 3.3}$$

and

$$A_c^{(0,p),-\infty}(X, \nu) = \text{Space of } \nu\text{-valuates compactly supported}$$

$$\mathit{(0,\ p)\text{-}\ forms\ with\ coefficients\ of\ generalized\ functions} \tag{X. 3.4}$$

Proof. [14].

We define the topological cohomology of Dolbeault de X, with values in ν, as

$$H_{top}^{0,p}(X, \nu) = [\ker \overline{\partial}](A^{0,p}(X, \nu) / \overline{\overline{\partial} A^{p-1,0}(X, \nu)}, \tag{X. 3.5}$$

Indeed, a quotient of the usual Dolbeault cohomology and carry a locally convex topology also usual. Similarly we define

$$H_{top}^{0,p}(X, \nu) = [\ker \overline{\partial}](A^{(0,p),-\infty}(X, \nu) / \overline{\overline{\partial} A_c^{(p-1,0),-\infty}(X, \nu)}, \tag{X. 3.6}$$

the topological cohomology of Dolbeault with compact supports. Then a natural identification exists

$$H^{0,p}_{top}(X,\mathfrak{I})^* \cong H^{0,n-p}_{C,top}(X,\Omega\otimes\mathfrak{I}^*), \qquad (X.\ 3.7)$$

Here $\mathfrak{I}^*$, is the vector holomorphic bundle dual of $\mathfrak{I}$.

We consider the following case. When X, is compact then the sub index c, is not added more, and the operators $\bar{\partial}$, automatically have closed range.

The central idea in this part of the program of Vogan is the desire to build representations of real reductive groups G, beginning with a measurable complex flag manifold X = G/L, and using G-equivariants holomorphic bundles of lines X. Indeed, if we have X = K/T, with r = 0, in $H_c^{0,r}(X,\nu)$, then $H_c^{0,r}(X,\nu)$, in the irreducible G-module on the corresponding coherent sheaf of $O(\lambda)$, global sections of the complex holomorphic bundle ν, and the relationship among $\bar{\partial}$ - cohomology and the sheaf is simple and it is given by the space

$$\Gamma(X,O(\lambda)) = H^0(X,O(\lambda)),$$

For the infinite case, is necessary to use a finer structure of the flag manifolds like for example, the given by open orbits of flag manifolds and the continuous homomorphisms among whose open orbits to induce a classification of the irreducible representations that reside in the space $H^0(X,O(\lambda))$, and that under the association of irreducible minimum K-types suggested by Vogan [38], the widest class in classifiable irreducible unitary representations will be obtained by the theory of Langlands.

However we will establish a special formalization of the $\bar{\partial}$ - cohomology to be able to use G-invariant holomorphic bundle of lines on X.

For compact G, the theorem of Borel-Weil says that all the irreducible representations of G, arise in this way as spaces of holomorphic sections of holomorphic bundles of lines.

Def. III. 2. Suppose that X, is a complex manifold of dimension *n,* and ν, is a holomorphic vector bundle on X. The cohomology of (p, q)-Dolbeault compactly supported of X, with coefficients in ν, is for definition

$$H_c^{0,p}(X,\nu)^* = (\ker(\bar{\partial})(A_c^{(p,q),-\infty}(\nu)) / (\operatorname{Im}(\bar{\partial}(A^{(p,q-1),-\infty}(\nu)), \qquad (X.\ 3.8)$$

If ν, is of finite dimension then the cohomology $H_c^{p,q}(X,\nu)$, is a cohomology of Čech with compact supports of X, with coefficients in rhe sheaf $\mathcal{O}_{\Omega_p\otimes\nu}$ of holomorphic p-forms with values inν.

Exactly a topology natural quotient exists on this cohomology, and we can define:

$H^{p,q}_{c,top}(X, v) =$ *Hausdorff maximal Quotient of*

$$H^{p,q}_{c}(X, v) = \mathrm{Ker}(\bar{\partial}) / \overline{\mathrm{Im}(\bar{\partial})}, \qquad (X.\ 3.9)$$

Then $H^{p,q}_{top}(X, v)$, it takes soft representations of G (for translation of forms).

Then clear consequence of the theorems of Serre, theorem X. 3. 2, and the theorem of Wong, theorem. X. 3. 1, using the definitions previous is have the corollary:

Corollary X. 3.1. (Bratten) [37]. Suppose that X, is a complex manifold G/L, and assume that admissible representation V is the minimal globalization of the ($\mathfrak{q}$, L∩K)-module. Be $A^{p,q}_{c}(X, v)$, the Dolbeault complex to v, with coefficients of generalized functions of compact support. Then the operator $\bar{\partial}$, have a closed range such that each one of the corresponding cohomological spaces $H^{p,q}_{c}(X, v)$, takes soft representation of G (on the dual of a nuclear Fréchet space). Everyone of this representations of G, is admissible and is a minimal globalization of their underlying module de Harish-Chandra.

Proof. (Vogan, 2000, Bratten, 2002).

The fundamental relation of duality of minimal and maximal globalizations is given by the corollary. X. 3.1, that makes allusion to the conjecture of Serre.

The theorem demonstrated by Bratten (Bratten, 2002), is different:

He defines a sheaf of germs of global sections $A(X, v)$, and it demonstrates a parallel result for the cohomology of sheaves with compact support on X, with coefficients in $A(X, v)$.

When V, is of finite dimension, the two results are exactly the same one, being this easy to verify for that Dolbeault cohomology (with coefficients of generalized functions of compact support) and it calculates the cohomology of sheaves in that case.

The case of infinite dimension of V, comparing the corollary one enunciated previously with results of Bratten is more difficult of establish. The development of Vogan (Complex Analysis and Unitary Representations, Springer, 2003) in all the exhibitions only speech of the Dolbeault cohomology and not of the cohomology of sheaves, foreseeing that the relationship between sheaves and the Dolbeault cohomology for bundles of infinite dimension is complicated and it bears bigger difficulties that those foreseen by the own theory of representations.

For the theory of characters, they are been able to determine the infinitesimal characters of the Dolbeault cohomology of representations and it is applicable the theorem of Vogan for representations of infinite dimension [37], [38].

The same way to the Dolbeault cohomology with compact support. The weight$\lambda_L - \rho(\mathfrak{u})$ that appears in the following corollary is therefore the infinitesimal character of the representation $H_c^{0,r}(X,\nu)$, with ν, having defined by $\nu \to X = G/L$.

Corollary X. 3.2. (Vogan). Be $Z = K / L \cap K$ a complex compact s-dimensional submanifold of the complex n-dimensional manifold X = G/L. Be $r = n - s$, the codimension of Z, in X. Assume that V is a ($\mathfrak{q}$, L)-module of infinitesimal character $\lambda_L \in \mathfrak{h}^*$, and that V, is the minimal globalization of the ($\mathfrak{q}$, L∩K)-module. Assume that $\lambda_L - \rho(\mathfrak{u})$, is weakly anti-dominant to $\mathfrak{u}$, this is that $-\lambda_L + \rho(\mathfrak{u})$, is weakly dominant. Then:

i. $H_c^{0,q}(X,\nu) = 0$, under $q = r$.

ii. Si $L = L_{max}$ and V, is a irreducible representation of L, then $H_c^{0,r}(X,\nu)$, is irreducible or cero.

iii. If the module of Harish-Chandra of V, admits a Hermitian and invariant form, then the module of Harish-Chandra of $H_c^{0,r}(X,\nu)$, admits a Hermitian form.

iv. If the module of Harish-Chandra of V, is unitary then the module of Harish-Chandra of $H_c^{0,r}(X,\nu)$, is unitary.

Proof. (Complex Analysis and Unitary Representations, Springer, 2003) and of the corollary X. 3. 1.

X. 4. $\mathfrak{u}$ — Cohomology

Be $\mathfrak{g}$, a reductive Lie algebra on ℂ. Be $\mathfrak{h}$, a cartan subalgebra of $\mathfrak{g}$, and be $\mathfrak{l}$, the system of positive roots to $\Phi(\mathfrak{g}, \mathfrak{h})$. Be $\mathfrak{b} = \mathfrak{b}(P) = \mathfrak{h} \oplus \oplus_{\alpha \in P} \mathfrak{g}_\alpha$. Be $\mathfrak{q}$, the subalgebra of $\mathfrak{b}$, enclosed $\mathfrak{b}$.

Be $\Phi_{\mathfrak{l}} = \{\alpha \in \Phi \mid (\mathfrak{g}_\alpha + \mathfrak{g}_{-\alpha}) \subset \mathfrak{q}\}$ and corresponding$\Sigma = P - \Phi_{\mathfrak{l}}$, with

$$\mathfrak{l} = \mathfrak{h} \oplus \oplus_{\alpha \in \Phi_{\mathfrak{l}}} \mathfrak{g}_\alpha, \qquad \text{(X. 4.1)}$$

and

$$\mathfrak{u} = \oplus_{\alpha \in \Sigma} \mathfrak{g}_\alpha, \qquad \text{(X. 4.2)}$$

Then $\mathfrak{q} = \mathfrak{l} \oplus \mathfrak{u}$ and $[\mathfrak{l},\mathfrak{u}] \subset \mathfrak{u}$. Be $\mathfrak{u}^- = \mathfrak{h} \oplus \oplus_{\alpha \in \Sigma} \mathfrak{g}_{-\alpha}$, and $\mathfrak{q}^- = \mathfrak{l} \oplus \mathfrak{u}^-$. Then $\mathfrak{q} = \mathfrak{u} \oplus \mathfrak{l} \oplus \mathfrak{u}^-$. By the theorem PBW (Poicaré-Bott-Weil),

$$U(\mathfrak{g}) = U(\mathfrak{l}) \oplus (\mathfrak{u}U(\mathfrak{g}) \oplus U(\mathfrak{g})\mathfrak{u}^-), \qquad \text{(X. 4.3)}$$

to $U(\mathfrak{g})$, enveloping algebra of Lie algebra $\mathfrak{g}$.

Be V, a $\mathfrak{g}$-module with actionπ. Then

$$C^i(\mathfrak{u}, V) = Hom_C(\wedge^i \mathfrak{u}, V), \tag{X. 4.4}$$

is a $\mathfrak{l}$-module under the action $(X\mu)(Y) = X(\mu(Y - \mu(adX(Y)))$, $\forall X \in \mathfrak{l}$, and $Y \in \mathfrak{u}$. Also $d(X\mu) = Xd\mu$.

The module given for (17) are the complexes of the $\mathfrak{u}$-cohomology.

X. 5. Generalized G-modules and ond theorem to representations of infinite dimension

G-modules of Fréchet are induced and irreducible G-modules of infinite dimension are built whose differentiable cohomology is a cohomology of representations of applicable infinite dimension to the Langlands classification and some geometric theorems as the theorem of Borel-Weil.

Def. X. 5.1. For a topological G-module or simply a G-module (π, V) will understand a topological vector space on which G acts via a continuous representation. A$\mathfrak{g}$-module is the corresponding pre-image of a G-module of the corresponding exponential homomorphism [37, 38, 46-48].

An extension of a G-module is a open G-orbit of a holomorphic bundle of flags on Fréchet spaces [46-48].

One generalization of the extension of a G-module $H^q(G(w), O_q(E_\eta)) \neq 0$, $\forall q \neq s$, is the case when η, is of infinite dimension. For this case is necessary to build a version of extension of G-module whose cohomology is the corresponding to a cohomology of representations of infinite dimension. s is the complex dimension of a compact maximal submanifold Z(w), of G(w), such that

$$Z(w) \cong K / K \cap L, \tag{X. 5.1}$$

and $E_\eta \to G(w)$, the homogeneous complex holomorphic vector bundle corresponding to the open G-orbit G(w) $\forall \eta \in \hat{L}$, and maximum weigh λ.

If L = T, the extension of the G-module is reduce to set of global sections of the sheaf $O_q(\lambda)$, of the complex holomorphic bundle of flag manifolds with maximum weigh λ. This is an irreducible G-module of finite dimension with maximum weigh λ (Theorem of Borel-Bott-Weil).

A version of extension of G-module whose cohomology is that of representations of infinite dimension can be built starting from the induction of G-modules on a differentiable cohomology defined as follows:

Def. X. 5.2. A generalized open G-orbit is the extension of G-module (open L-module on a differentiable cohomology) induced in the differentiable category given by the space

$$I^{\infty}(G(w)) = \mathrm{Ind}_{L}^{G}G(w) \ [11], \qquad (X.\ 5.2)$$

Def. IV.3. (Bulnes, F.) [37, 38, 46-48]. A generalized G-module is the induced G-module by a differentiable cohomology of representations of infinite dimension (E,η), defined on generalized orbits of the complex homogeneous bundle

$$E_{\eta} \to G/L \ [8] \qquad (X.\ 5.3)$$

where E, is a Fréchet space.

Using $\mathfrak{u}$-cohomology, continuous cohomology and the generalization of the topology on complexes of fibered holomorphic bundles of Frèchet is having that:

Proposition X. 5.1. (Bulnes, F.) [37, 38].

$$H^{\bullet}_{ct}(\mathfrak{u}, I^{\infty}(\eta)) = H^{\bullet}_{ct}(G/L, \mathcal{O}_{\mathfrak{q}}(E_{\eta})), \qquad (X.\ 5.\ 4)$$

Proof. [38].

Of these generalities in hand, we get immediately a description of the topological dual of Dolbeault cohomology.

Using the conjecture of Vogan on the possible extension of representations of L to modules of Harish-Chandra to G taking care that the co-border operators of the Dolbeault cohomology have all closed range, we apply this conjecture on the extension of the induced G-modules proposed to modules of Harish-Chandra of infinite dimension, obtaining the conjecture:

Conjecture X. 5.1. (Bulnes, F.) [38]. Suppose that $H^{q}_{ct}(\mathfrak{g}, L\cap K; A^{L}\otimes E_{\gamma}^{*})$ is a $(\mathfrak{l}, L\cap K)$-module of finite longitude, E_{η}, their corresponding representation of L, andv, the make holomorphic associated bundle to G/L.Then the co-border operators of the Dolbeault cohomology are all of closed range and to the case of infinite dimension the range of the co-border operators of the Dolbeault cohomologyare closed provided certain intertwining operators [39], applied to the corresponding induced G-modules are modules of Harish-Chandra (which must satisfy the theorem of Vogan-Zuckerman on irreducible unitary representations of infinite dimension under character of Vogan).

Using certain technical lemmas [10], to a decomposition of the algebra $\mathfrak{u}$ in their parts extensive y classified of the radical nilpotent part of the complex holomorphic vector bundle module a radical nilpotent bundle of Borel subalgebras [8, 12], and with pertinent generalizations of the G-modules, is have a theorem of classification of representations of infinite dimension [8]:

Theorem X. 5.1. (Bulnes, F.)[21, 37, 38, 41, 46, 47], and [48]. Be η, denotes for $H^{t}_{ct}(L(w), O_{\mathfrak{q}}(E_{\gamma v}))$, a representation of infinite dimension of ($\mathfrak{q}$, L) and be $E_{\eta} \rightarrow G(w)$, the associated homogeneous vector bundle. Then the operator $\bar{\partial}$, to the complex de Dolbeault $A(G(w), E_{\eta}\otimes(\wedge^{q+1}\mathfrak{u})^{*})^{L}$, is of closed range. Then the cohomologies $H^{q}(G(w), O_{\mathfrak{q}}(E_{\eta})) = 0$, $\forall$ $q \neq s$ are admissible G-modules de Frèchet composition of series (These form admissible representations of finite longitude). Their modules of Harish-Chandra underlying are functors of Zuckerman [21, 48] $A^{s+t}(G, M, \mathfrak{b}, \gamma_v) = A(G, L, \mathfrak{q}, \eta)$. E_{η}, have Infinitesimal Character $\eta_{L,\lambda}$, and trivial action $\mathfrak{u}$ (to it the generalized G-modules admit the infinitesimal G-character $\gamma_{vG,\lambda+\rho(\mathfrak{u})}$).

Proof. To that $\bar{\partial}$, be close is necessary that be regular in whole their domain (Wong's globalizations). Then $\bar{\partial}$, on $H^{q}(G(w), O_{q}(E_{\eta}))$, is a representative K-finite cohomology of strong harmonic L^2-form. By the intertwining operator [38, 39], it is map fundamental series in harmonic forms of I(w), of $H^{q}(G/L, \mathcal{L}_{\gamma v})$, with $\mathcal{L}_{\gamma v}$, the bundle of lines and γ_v, is the unitary character of $L \subset G$. Then $\bar{\partial}$, is regular in $A^{q}(G, L, \mathfrak{q}, \eta)$. Then by differentiable cohomology, $rS(F_0)$. Then by differentiable cohomology

$$\rho S(\Phi_0) \neq 0, \tag{X. 5.5}$$

$\forall$ $F_0 \subset Q(F)$ with *K*-type (μ_0, F_0) in $C^{\infty}(L\cap G, \wedge^{q}\mathfrak{u} \otimes \mathbf{C}_{\gamma v})$, and $S(F_0)$, represent the class of cohomology non vanishing on $K \cap L$. Using results of $\mathfrak{u}$ – cohomology (lemmas of Vogan and Kostant of $\mathfrak{u}$ – cohomology) to $\mathfrak{g}$-modules. This class of cohomology is the of the ($\mathfrak{g}$, K) – modules isomorphic to space $H^{q}(G(w), E_{\eta}\sim)$, when $E_{\eta}\sim = \mathbf{C}_{\gamma v}$. But $\mathbf{C}_{\gamma v} = \mathbf{C}_{\rho(\mathfrak{u})+\lambda}$, and due to that the induced representations on generalized G-modules like the defined in the Def. X. 5. 1, and Def. X. 5. 2, can be identified under the Szegö intertwining operator

$$S: Ind_{\mu}^{L}(E_{\sigma}\otimes\rho_{L\otimes 1}) = E_{\gamma v} \rightarrow H^{q}(G/L, E_{\gamma v}), \tag{X. 5.6}$$

where

$$H^{q}(G/L, E_{\gamma v}) \hookrightarrow Ind_{L}^{G}(E_{\eta}\sim \otimes (\wedge^{q+1}\mathfrak{u})^{*})), \tag{X. 5.7}$$

and $Ind_{L}^{G}(E_{\eta}\sim \otimes (\wedge^{q+1}\mathfrak{u})^{*})) = E_{\eta}$, where

$$E_{\eta} \rightarrow H^{q}(G(w), O_{\mathfrak{q}}(E_{\eta})), \tag{X. 5.8}$$

Not there is to lose of see that it is wants to carry on the classical representations $Ind_{\mu}^{L}(E_{\sigma}\otimes\rho_{L\otimes 1})$, with discrete series σ, built up by the Vogan's algorithm of minimal K-types of $A_{\mathfrak{q}}(\lambda)$ to the canonical temperate representations $Ind_{L}^{G}(E_{\gamma v}\otimes(\wedge^{q+1}\mathfrak{u})^{*}))$, where the restriction fiber of $E_{\gamma v}$, have Dolbeault operator $\bar{\partial}_{L(w)}$. In particular and using the generalization of the Borel-Bott-Weil theorem (to L, locally compact),

$$H^{0,q}(K/(K\cap L), \mathbf{C}_{\rho(\mathfrak{u})+\lambda}) = H^{0,q}(K/(K\cap L), \mathbf{C}_{2\delta(\mathfrak{u})+\lambda}) \cong F^{2\delta(\mathfrak{u})+\lambda}, \tag{X. 5.9}$$

We can to use these representations like complexes of Dolbeault to a globalization of Wong of type $C^{-\omega}(L(w), E_{\gamma\nu} \otimes (\wedge^{*}L_{L(w)})^{*})$.If we inducing certain representations $K(w) \cong K/(K \cap L)$, of G(w), to $L(w) \cong G/L$, of G(w). But this is possible due to the construction of the generalized G-modules. Then $\overline{\partial}$, on whose complexes is close.

Then using likeu = $\dim_{\mathbb{C}}(K \cap L)(x)$, with u = t + s, with t, and s, complex dimensions of $(K \cap M)(x)$, and $(K \cap L)(w)$, and the fact of that

$$H^{q}(M(x), \mathfrak{o}_{\mathfrak{b}}\sim(E_{\gamma\nu\sim})) = 0, \qquad \text{(X. 5.10)}$$

$\forall\, p \neq t$, and

$$H^{q}(L(x), \mathfrak{o}_{\mathfrak{b}}(E_{\gamma\nu})) = 0, \qquad \text{(X. 5.11)}$$

$\forall\, p \neq u$, we have the spectral succession of Leray of $L(x) \to L(w)$, collapse to E_2. Then the vanishing of the group of cohomology $H^{q}(L(x), \mathfrak{o}_{\mathfrak{b}}(E_{\gamma\nu})) = 0$, $\forall \lambda + \rho(\mathfrak{u})$, establish that

$$H^{s+t}(L(x), \mathfrak{o}_{\mathfrak{b}}(E_{\gamma\nu})) = H^{s}(L(x), \mathfrak{o}_{\mathfrak{q}}(H^{t}(M(x), \mathfrak{o}_{\mathfrak{b}}\sim(E_{\gamma\nu\sim})))), \qquad \text{(X. 5.12)}$$

and due to the development of the relative lemmas of Frèchet spaces it is follows that (X. 5. 12) have structure of Frèchet space to the which the action of G, is a continuous representation (generalized G-module).□

Cohomologically should be induced with finer decompositions of an nilpotent algebra of $\mathfrak{u}$, obtaining spaces more classified "thin"; the corresponding admissible G-modules of Frèchet. The obtained theorem can classify a great part of representations of infinite dimension although not their entirety due to the difficulty of obtaining a substantial algebra whose Dolbeault cohomology has an operator of closed range on the admissible G-modules that appear in the problem of representation of those ($\mathfrak{l}$, $L \cap K$)-modules for an algebra $\mathfrak{q}$.

However, choosing an appropriate infinitesimal G-character, we could establish functors of Zuckerman corresponding to a algebra $\wedge^{p}\mathfrak{u}$, with $\mathfrak{u} = \mathfrak{g}/\mathfrak{t}$, and canonical globalizations X^{g}, with $g = \infty$, $g = \omega$, $g = -\infty$, or $g = -\omega$; having the property of closed range.

The election of an appropriate infinitesimal G-character causes redundancies in the admissible G - modules that turn out to be unitary representations, for what is important to choose a decomposition of u in the group of Levi.

APPENDIX A: Integral Formulas on Canonical p-Pairs

Be G, a real reductive group. We fix θ, a involution of Cartan G = NAK, and F, the subset of Δ_0, with $\Delta_0 = (\Delta - F_0)|_{\mathfrak{a}}$, where $F_0 = \{\alpha \in \Delta \mid \alpha|_{\mathfrak{a}} = 0\}$, and Δ is the system of positive simple roots of $\Phi(\mathfrak{g}, \mathfrak{a})$. Be (P_F, A_F), the canonical p-pair. If $\mu \in (\mathfrak{a}_F)^*$, and if $H \in \mathfrak{a}_F$, then $a^{\mu} = \exp\mu(H)$, if a = exp H. We define to $\rho_F \in (\mathfrak{a}_F)^*$, by

$$\rho_F(H) = (1/2)\mathrm{tr}(\mathrm{ad}\, H|_{\mathfrak{n}_F}). \quad \text{(A. 1)}$$

Lemma. A. 1. Let dn, da, dm, be invariant measures on N_F, A_F, ${}^0M_F$. Let dk, be a normalized invariant measure on K. Then we can to elect an invariant measure dg, on G, such that

$$\int_G f(g)\, dg = \int_{N_F \times A_F \times {}^0M_F \times K_F} f(namk) a^{-2\rho_F}\, dn\, da\, dm\, dk, \quad \text{(A. 2)}$$

to $f \in C_c(G)$. Also if $u \in C(K)$, then

$$\int_K u(k)\, dk = \int_{K \times K_F} u(k_F k(kg)) a(kg)^{2\rho_F}\, dk_F\, dk, \quad \text{(A. 3)}$$

where if g∈G, and if g = nak, n∈N, a∈A, and k∈K, then a(g) = a, and k(g) = k.

Proof. Let dp, be a left invariant measure on P_F. Then we can elect an invariant measure, dg, on G, such that

$$\int_G f(g)\, dg = \int_{P_F \times K} f(pk)\, dp dk, \quad \text{(A. 4)}$$

By the *lemma. I. 2.2.* For which can be demonstrated that save a scalar multiple $dp = a^{-2\rho_F}$ dndadm. The lemma on suprajective diffeomorphisms[9] implies that dp = h(n, a, m)dndadm, with h, a smooth function on $N_F \times A_F \times {}^0M$. By left invariance h, is independient of n.

By definition of ${}^0M_F$, the modular function, δ, of P_F, is 1, on ${}^0M_F$. Thus dp, is right invariant under ${}^0M_F$. Of where h, is a function of only one of the components.

The Jacobian of the action $n \mapsto ana^{-1}$ is $\det(\mathrm{Ad}(a)|_{\mathfrak{n}}) = a^{2\rho_F}$, to $a \in A_F$. Thus $a^{-2\rho_F}$dndadm, is left A_F_invariant.

[9] **Lemma.** (1). The map $M_F \times N_F \to P_F$, given by m, n $\mapsto$ mn, is a suprajective diffeomorphism.
(2). The map ${}^0M_F \times A_F \times N_F \to P_F$, given for m, a, n $\mapsto$ man, is a suprajective diffeomorphism.

Now we demostrate the second afirmation of the lemma. Remembering the lemma. I. 2.1., exist a continuous function compactly supported f, on G, such that

$$\int_{P_F} f(pk)\, dp = \int_{K_F} u(k_F k)\, dk_F, \quad \forall\ k \in K \tag{A. 5}$$

Thus have

$$\int_G f(x)\, dx = \int_K u(k)\, dk, \tag{A. 6}$$

Now,

$$\int_G f(x)\, dx = \int_G f(xg)\, dx = \int_{P_F \times K} f(pkg)\, dp\, dk, \tag{A. 7}$$

Writing $kg = na(kg)k(kg)$, like above, dp is transformed for δ, under right multiplication by elements of P_F. Since

$$\delta(na(kg)) = a(kg)^{2\rho_F},$$

we have

$$\int_K u(k)\, dk = \int_{P_F \times K} a(kg)^{2\rho_F} f(pk(kg))\, dp\, dk = \int_{K_F \times K} u(k_F k(kg)) dk_F\, dk,$$

that is the wanted result. ■

To the following formula of integration we asume that G, is of inner type. Let R, be a system of positive roots to $\Phi(\mathfrak{g}, \mathfrak{a})$, corresponding to the election of $\mathfrak{n}$. Be $\mathfrak{a}^+$, equal to the corresponding Weyl camera to R[10].

Be $A^+ = \exp(\mathfrak{a}^+)$. If $a \in A$, $a = \exp H$, we do corresponding

$$\gamma(a) = \prod_{\alpha \in R} \sinh(\alpha(H)). \tag{A. 8}$$

Lemma. A. 2. dg, can be normalized such that

$$\int_G f(g)\, dg = \int_{K \times A^+ \times K} \gamma(a) f(k_1 a k_2)\, dk_1 da\, dk_2. \tag{A. 9}$$

---

[10] Be $\mu \in P$. Be $X \in \mathfrak{g}^\mu$, such that $\langle X, X \rangle = 1$. Then $[X, \theta X] = -H_\mu$. Here $H_\mu \in \mathfrak{a}$, is defined by $B(H, H_\mu) = \mu(H)\ \forall\ H \in \mathfrak{a}$. Thus if $x = (2/\mu(H_\mu))X$, $y = -\theta X$, $h = (2/\mu(H_\mu))H_\mu$, then x, y, h generate a TDS (three Dimensional Simple) Lie algebra on **IR**. Thus exist a homomorphism de Lie, σ, of SL(2, **IR**), in G_0, such that $\sigma(g^*) = (\theta\sigma(g))^{-1}$. Be k, the image of $[0\ \ 1 \mid -1\ 0]$, under σ. Then, if s_μ, is defined by $s_\mu H = H - B(h, H)H$, to $H \in \mathfrak{a}$, then $Ad(k)H = s_\mu H$. Be $N(\mathfrak{a}) = \{u \in K^0 \mid Ad(u)\mathfrak{a} = \mathfrak{a}\}$.
Be $W(\mathfrak{g}, \mathfrak{a}) = \{Ad(u)|_{\mathfrak{a}} \mid u \in N(\mathfrak{a})\}$. Then $s_\mu \in W(\mathfrak{g}, \mathfrak{a})$, to all $\mu \in P$. This it is follows of the realized observations with before into of this foot of page.
Be $\mathfrak{a}'$, the set of all the $H \in \mathfrak{a}$, such that $\mu(H)$, is not null to all $\mu \in \Phi$. A connect component of $\mathfrak{a}'$, is called a Weyl camera of $\mathfrak{a}$. If C, is a Weyl camera then the set of all the $\mu \in \Phi$, such that μ, is positive on C, denoted by the space P_C, is called a positive roots systems. If $\mu \in P$, and if μ, can not be write like a sum of two elements of P, then μ, is called simple in P. One interesting proposition to respect is:
Proposition. (1) $W(\mathfrak{g}, \mathfrak{a})$ is generate by the s_μ, to μ, simple in P.
(2) $W(\mathfrak{g}, \mathfrak{a})$ act simple and transitive on the Weyl camera of $\mathfrak{a}$.

Proof. Let

$$\beta : \mathfrak{a}^{+} \times K/M^{0} \rightarrow \mathfrak{p}, \tag{A. 10}$$

defined by $\beta(H, kM^{0}) = Ad(k)H$. Be $\mathfrak{p}'$, the rank of β. Since $Ad(K)\mathfrak{a} = \mathfrak{p}$, $Ad(K)\mathfrak{a}^{+} = Ad(K)\mathfrak{a}'$, and $Ad(K)(\mathfrak{a} - \mathfrak{a}')$, is a finite union of submanifolds of low dimension, $\mathfrak{p}'$, is open, dense and have a complement of measure 0, in $\mathfrak{p}$. Is easy demonstrate that β, is a diffeomorphism in $\mathfrak{p}'$(see proposition (1), foot of page 2).

Let P_j, be a system of positive roots for $\Phi(\mathfrak{g}_C, (\mathfrak{h}_j)_C)$. Set $\pi_j(H) = \Pi_{\alpha \in P}\alpha(H)$, for $H \in \mathfrak{h}_j$. Let D, is a non-zero polynomial function on $\mathfrak{g}$. Then $|D(H)| = |\pi_j(H)|^2$. Since G, and each H_j, are unimodular, each coset space G/H_j, has a G-invariant measure dx_j.

Proposition. A. 1. There exist positive constants c_j, $j = 1, \ldots, r$, and normalizations of Lebesgue measure on $\mathfrak{g}$, and the $\mathfrak{h}_j$, such that

$$\int_{\mathfrak{g}} f(X)dX = \sum c_j \int_{\mathfrak{h}_j} |D(H)| \left(\int_{G/H_j} f(Ad\, xH)dx_j \right) dH, \tag{A. 11}$$

for $f \in C_c(\mathfrak{g})$.

Proof. For the moment, fix j, and let $\mathfrak{h}_j = \mathfrak{h}$, etc. Let μ: $G/H \times \mathfrak{h}' \rightarrow \mathfrak{g}$, be defined by $\mu(gH, h) = Ad(g)h$ (here $\mathfrak{h}' = \mathfrak{g}' \cap \mathfrak{h}$). We may identity the complex tangent space at 1H, to G/H, with $\mathfrak{n}^{+} + \mathfrak{n}^{-}$. Translating by the elements of G, allows us to identity the tangent space at gH, with this space. A direct calculation yields

$$d\mu_{gH,\, h}(X, Z) = Ad(g)(ad\, Xh + Z), \tag{A. 12}$$

for $X \in \mathfrak{n}^{+} + \mathfrak{n}^{-}$, $Z \in \mathfrak{h}$. This implies

i. The Jacobian of μ, at gH, h, is D(h), up to sign.

This implies that μ, is everywhere regular. The remarks preceding the statement we are proving now imply that μ, is a [W]-fold covering of their range. Lemma[11] implies that $\mathfrak{g}'$, is the disjoint union of the open subsets $Ad(G)(\mathfrak{h}_j)_C$. The result now follows from (i).

The above result is sometimes called the Weyl integral formula for $\mathfrak{g}$.

Now we derive the Weyl integral formula for G. We define real analytic functions d_j, on G by

$$\det(tI - (Ad(g) - I)) = \Sigma t^{j}\, d_j(g),$$

[11] **Lemma.** (1) If $X \in \mathfrak{g}'$, then X, is semi-simple and $C_{\mathfrak{g}}(X) = \{Y \in \mathfrak{g} \mid [X, Y] = 0\}$, is a Cartan subalgebra of $\mathfrak{g}$.
(2) If X, is a semi-simple element of $\mathfrak{g}$, then $C_{\mathfrak{g}}(X)$, is a reductive subalgebra of $\mathfrak{g}$, that contains a Cartan subalgebra.

Here n = dimG. Set d = d_j, for j = rank($\mathfrak{g}_c$). We set G′ = {g∈G | d(g) ≠ 0}. Then G′, is open, dense with complement of measure 0, in G.

Proposition A. 2. There exist positive constants m_j, so that if dg and dh_j, are respectively invariant measure on G, and H_j, then

$$\int_G f(g)dg = \sum m_j \int_{H_j} |d(h_j)| \left(\int_{G/H_j} f(gh_j g^{-1}) d(gH_j) \right) dh_j, \quad \forall \; f \in C_c(G), \tag{A. 13}$$

Proof. We fix j, and for the moment drop the index j. Let σ:G/H × H′ → G, be defined by σ(gH, h) = ghg^{-1}(here H′ = H ∩ G′). We have

$$d\sigma_{gH,h}(X, Z) = (Ad(g)((Ad(h^{-1}) - I)X + Z))_{\sigma(gH,h)}, \tag{A. 14}$$

for $X \in \mathfrak{n}^+ \oplus \mathfrak{n}^-$, $Z \in \mathfrak{h}$. The rest of the proof is now almost identical to that of proposition A. 1, and the details can be done like a exercise.

A result that helps us to derive some integration formulas that are related to the Gelfand-Naimark decomposition is the following result.

We set $V_F = \theta N_F$. Fix invariant measures dn, dm, da, dv, respectively on N_F, ^0M_F, A_F, and V_F.

Lemma A. 2. The invariant measure dg, can be normalized so that

$$\int_G f(g)dg = \int_{N_F \times {}^0M_F \times A_F \times V_F} a^{-2\rho_F} f(nmav)dndmdadv, \tag{A. 15}$$

for $f \in C_c(G)$. If $u \in C(K)$, then

$$\int_K u(k)dk = \int_{K_F \times V} a^{-2\rho_F} f(nmav)dndmdadv, \quad \text{(A. 16}$$

Proof. [4-6]. ■

APPENDIX B: Regularity on a Lie Algebra

One of the fundamental theorems that gives place to born of the ordinary D-modules is the theorem or regularity of Harish-Chandra. In this is established a discussion on involutive distributions require in a decomposition of Cartan of an open G-invariant of a reductive algebra $\mathfrak{g}$, that which permit to obtain germs of a sheaf of differential operators in a Fréchet space. This give place to construct a regularity theory of the (g, K)-modules through of differential operators (the classical D-modules) and their extensions on G-invariant folds of a differential manifold with subjacent reductive group G.

Let G, be a real reductive group, $\mathfrak{g}_0$, their real Lie algebra and $\mathfrak{g} = \mathfrak{g}_0 \otimes_{\mathbb{R}} \mathbf{C}$, their complexification, G^0, the topological component of the unit of G, defined as the space

$$G^0 = \{g\in G|Ad(g) = I, \forall\ Ad\in End(G)\}, \qquad (B. 1)$$

Given that G is reductive then Int($\mathfrak{g}$), is the space of automorphisms

$$Int(\mathfrak{g}) = \{g\in Aut(G)|exp(adX) = g, \forall\ X\in\mathfrak{g}\}, \qquad (B. 2)$$

where Int($\mathfrak{g}$) can indentifies like the space $Ad(G^0) \otimes_{\mathbb{R}} \mathbf{C}$.

If $X\in\mathfrak{g}$, and $g\in G$, then $Ad(g)X = gX$. If Ω, is an open G-invariant of $\mathfrak{g}$, and if $f\in C^{\infty}(\Omega)$, then $\tau(g)f(X) = f(g^{-1}X)$, $\forall\ \gamma\in G$, and $X\in\mathfrak{g}$. Let be the open space or open of operator invariants

$$D'(\Omega)^G = \{T\in D'(\Omega)|\ T\tau(g) = T, \forall\ g\in G\}, \qquad (B. 3)$$

Let $D = D_r(X)$, be with $r = \dim \mathfrak{h}$, with $\mathfrak{h} \subset \mathfrak{g}$, and such that $X\in\mathfrak{g}'$, where

$$\mathfrak{g}' = \{X\in\mathfrak{g}|\ D(X) \neq 0\}, \qquad (B. 4)$$

We consider $\Omega' = \Omega \cap \mathfrak{g}'$. Be $\mathfrak{h}_1, \ldots, \mathfrak{h}_s$, a complete set of Cartan subalgebras non-conjugates of $\mathfrak{g}$. Be $\Omega_j' = G(\Omega' \cap \mathfrak{h}_j)$, (Is the corresponding open orbit of the group G corresponding to the subalgebra of Cartan $\mathfrak{h}_j$). Then $\Omega' = \cup\Omega_j'$. Indeed, be $H\in\mathfrak{h}_j (j \leq s)$, then for the G-invariance of Ω,

$$Ad(g)H = gH, \qquad (B. 5)$$

where $gH\in G(\Omega' \cap \mathfrak{h}_j) = G[(\Omega \cap \mathfrak{g}') \cap \mathfrak{h}_j] = \Omega_j'$. Then

$$\Sigma_{j \leq s} Ad(G)H_j = \Sigma_{j \leq s} H_j = \mathfrak{g}', \forall\ g\in G, H\in\mathfrak{h}_j, \qquad (B. 6)$$

which is translated in $\cup_{j\leq s} G(\Omega' \cap \mathfrak{h}_j) = \cup_{j\leq s} \Omega_j' = \Omega'$. If $g\in G$ and if $H\in\Omega' \cap \mathfrak{h}_j$, then we define the map

$$\psi_j : G \times (\Omega' \cap \mathfrak{h}_j) \rightarrow \Omega_j', \tag{B. 7}$$

with rule of correspondence

$$(g, H) \mapsto gH, \tag{B. 8}$$

This map is a submersion of $G \times (\Omega' \cap \mathfrak{h}_j)$ in Ω_j'. That's right, by appendix A, if $H\in\Omega' \cap \mathfrak{h}_j$, with $g\in V$, and $Y\in\mathfrak{h}$, then

$$d\psi_j(X, Y) = \mathrm{Ad}(g)([X, H] + Y), \tag{B. 9}$$

where $d\psi_j$, is suprajective $\forall\ g\in G$, $H\in\Omega'\cap\mathfrak{h}_j$. Of where ψ_j, is submersion.

We fix the Lebesgue measures dX, and dH, on $\mathfrak{g}$, and each $\mathfrak{h}_j$, respectively. We visual to $S(\mathfrak{g})$, as the algebra

$$S(\mathfrak{g}) = \{D_X\in DO(\mathfrak{g})|\ D_X = \Sigma P_I(X)\partial^I,\ \forall\ X\in\mathfrak{g}\}^{12}, \tag{B. 10}$$

and consider the differential ideal $I(\mathfrak{g}) = S(\mathfrak{g})^G$, this due to that in $S(\mathfrak{g})^G$, we can establish a formula of differential ideals with the semi-simple structure of $\mathfrak{g}$.

We fix Cartan subalgebra $\mathfrak{h}$ of $\mathfrak{g}$. Be $\Phi = \Phi(\mathfrak{g}_C, \mathfrak{h}_C)$. If $\alpha\in\Phi$, and if $\alpha(\mathfrak{h}) \subset \mathrm{IR}$ (respectively $\alpha(\mathfrak{h}) \subset i\mathrm{IR}$), then we say that α, is real or imaginary respectively.

Let Φ_{IR}, and Φ_I, be the systems of real and imaginary roots respectively. Let $\Gamma_{\mathrm{IR}} = \Phi_{\mathrm{IR}}$, and $\Gamma_I = i\Phi_I$, be and $\Gamma = \Gamma_{\mathrm{IR}} + \Gamma_I$.

Let

$$\mathfrak{h}'' = \{H\in\mathfrak{h}|\alpha(H) \neq 0,\ \forall\ \alpha\in\Gamma\}, \tag{B. 11}$$

Clearly $\mathfrak{h}'' \supset \mathfrak{h}'$. In effect, if $H\in\mathfrak{h}'$, such that

$$\mathfrak{h}' = \{H\in\mathfrak{g}'|\alpha(H) \neq 0,\ \forall\ \alpha\in\Gamma\}, \tag{B. 12}$$

then $H\in\mathfrak{h}''$.

---

12 Sea $X_1, \ldots, X_n$, a base of $\mathfrak{g}$, and be $x_1, \ldots, x_n$, the corresponding coordinates of $\mathfrak{g}$. If $D\in DO(\mathfrak{g})$, then

$$D = \Sigma p_I\partial^I,$$

where has been used the notation of multi-indices ordinary. If $I = (i_1, \ldots, i_n)$, with i_n, a non-negative integer then $|I| = \Sigma i_j$, and $\partial^{|I|}/x^{i_1}{}_1, \ldots, x^{i_n}{}_n$. If $X\in\mathfrak{g}$, then $D_X = \Sigma p_I(X)\partial^I$. Then D_X, is a constant coefficient differential operator on $\mathfrak{g}$. Clearly, $T(I \otimes S(\mathfrak{g}_C))$ (to all homeomorphism of Lie algebras belonging to space $\mathrm{Hom}_{\mathfrak{g},K}(\mathfrak{l}, DO(\mathfrak{g}))$, where $\mathfrak{l}$, is the subalgebra of $\mathfrak{g}$ corresponding to Lie group $L = G \times \mathfrak{g}$, is the algebra of all the constant coefficient differential operators on $\mathfrak{g}$. We can thus identify a $S(\mathfrak{g}_C)$, with the algebra of constant coefficient differential operators on $\mathfrak{g}$.
Note: T, is extended to a homomorphism of algebras of $U(\mathfrak{l}_C)$, in $DO(\mathfrak{g})$.

Lemma. B. 1. Let C, be a connect component of $\mathfrak{h}''$. Then exist $\gamma_1, \ldots, \gamma_q \in \Gamma$, such that:

1. $\gamma_1, \ldots, \gamma_q$, are linearly independents,
2. $C = \{H \in \mathfrak{h} | \gamma_j(H) > 0, j = 1, 2, \ldots, q\}$.

As consequence $C \cap \mathfrak{h}'$, is connecting.

Proof. If $\alpha \in \Gamma$, then $\alpha(z(\mathfrak{g})) = 0$, since

$$z(\mathfrak{g}) = \{X \in \mathfrak{g} | \alpha(X) = 0\}, \tag{B. 13}$$

Thus we can assume unloosed of generality that $\mathfrak{g}$, is semi-simple. Let

$$\mathfrak{h}_{IR} = \{H \in \mathfrak{h}_C | \alpha(H) \in IR, \forall\ \alpha \in \Phi^{+}\}, \tag{B. 14}$$

Then $\mathfrak{h} = \mathfrak{h}_{IR} \oplus i\mathfrak{h}_{IR} = (\mathfrak{h}_{IR} \cap \mathfrak{h}) \oplus (i\mathfrak{h}_{IR} \cap \mathfrak{h})$. If $\alpha \in \Gamma_{IR}$, (or $\alpha \in \Gamma_I$, respectively) then $\alpha(\mathfrak{h}_{IR} \cap \mathfrak{h})$ (or $\alpha(i\mathfrak{h}_{IR} \cap \mathfrak{h}) = 0$).

Consider the components spaces

$$(\mathfrak{h}_{IR} \cap \mathfrak{h})' = \{H \in (\mathfrak{h}_{IR} \cap \mathfrak{h})\ |\alpha(H) \neq 0, \forall\ \alpha \in \Gamma_{IR}\}, \tag{B. 15}$$

and

$$(i\mathfrak{h}_{IR} \cap \mathfrak{h})' = \{H \in (i\mathfrak{h}_{IR} \cap \mathfrak{h})\ |\alpha(H) \neq 0, \forall\ \alpha \in \Gamma_I\}, \tag{B. 16}$$

Then a connect component of $\mathfrak{h}''$, is the space of Weyl cameras $C_1 \times C_2$, with $C_1 \cap (\mathfrak{h}_{IR} \cap \mathfrak{h})'$, and $C_2 \cap (i\mathfrak{h}_{IR} \cap \mathfrak{h})'$, connect. Given that, Γ_{IR}, and Γ_I, are both systems of roots and $\mathfrak{h}$, a Cartan subalgebra of $\mathfrak{g}$, then $\Gamma = \Gamma_{IR} \cup \Gamma_I$, is a system of roots corresponding to the Weyl cameras of $C \cap \mathfrak{h}'$. Then $\gamma_1, \ldots, \gamma_q$, result linearly independents. Thus if we consider

$$\Sigma = \Phi - (\Gamma_{IR} \cap \Gamma_I), \tag{B. 17}$$

and if $\alpha \in \Sigma$, then $Re\alpha$, and $Im\alpha$, are linearly independents. Therefore the root spaces of $\mathfrak{h}$, relative to α, is

$$(\mathfrak{h}_j)_\alpha = \{H \in \mathfrak{h}_j | \alpha(X) = 0\}, \tag{B. 18}$$

is such that $codim(\mathfrak{h}_j)_\alpha = 2$ in $\mathfrak{h}_j$. Thus in $C \cap (\mathfrak{h}_j)'$, $(\mathfrak{h}_j)_\alpha$, have codimension

$$codim(\mathfrak{h}_j)_\alpha = dim\ C - dim \cup_{\alpha \in \Sigma} (\mathfrak{h}_j)_\alpha, \tag{B. 19}$$

Of which the space $C \cap (\mathfrak{h}_j)' = C - \cup_{\alpha \in \Sigma} (\mathfrak{h}_j)_\alpha$, is connect.

Consider to Ad(G), such that

$$Ad(G)Z(\mathfrak{g}) = Z(\mathfrak{g}), \tag{B. 20}$$

(Act trivially on the center of the Lie algebra of Lie $\mathfrak{g}$).

Let $\phi_1, \ldots, \phi_d$, be homogeneous Ad(G)-invariants polynomials on $[\mathfrak{g}, \mathfrak{g}]$, (that is to say, $\phi_j = X_j Y_j - Y_j X_j$) ($j = 1, \ldots, d$), such that

$$[Ad(g)X_j, Ad(g)Y_j] = [X_j, Y_j], \tag{B. 21}$$

Then $\forall\, r > 0$,

$$\Omega(\phi_1, \ldots, \phi_d, r) = \{X \in [\mathfrak{g}, \mathfrak{g}] \,|\, |\phi_j(X)| < r, j = 1, \ldots, d\}, \tag{B. 22}$$

Let U, be an open and connect subset of $Z(\mathfrak{g}) = Z$. Let

$$\Omega = \{X + Y \,|\, X \in U, \text{ and } Y \in \Omega(\phi_1, \ldots, \phi_d, r)\}, \tag{B. 23}$$

Lemma. B. 2. Ω is connect. More yet, if $\mathfrak{h}$, is a Cartan subalgebra of $\mathfrak{g}$, and if C is a connect component of $\mathfrak{h}'$, then $C \cap \Omega$, is connect.

Proof. If $X \in U$, and $Y \in \Omega(\phi_1, \ldots, \phi_d, r)$, then $X + tY \in \Omega$, to $0 \leq t \leq 1$. Then implies by connectivity that Ω, is connect. To demonstrate the second affirmation is sufficient consider to $\mathfrak{g} = [\mathfrak{g}, \mathfrak{g}]$, due to the definition of the space $\Omega(\phi_1, \ldots, \phi_d, r)$. Indeed, considering to B a connect neighborhood of the 0, such that $B_e \subset C \cap \Omega$, and C, a connect component of $\mathfrak{h}''$, where

$$\mathfrak{h}'' = \{H \in \mathfrak{h} \,|\, \alpha(H) \neq 0, \forall\, \alpha \in \Gamma\},$$

where $\Gamma_{IR} = \Phi_{IR}$, and $\Gamma_I = \Phi_I$, and $\Gamma = \Gamma_{IR} \cup \Gamma_I$. By the lemma I. 1, $C \cap \mathfrak{h}'$, is connect. Since $\mathfrak{h}'' \supset \mathfrak{h}'$, then C like component of $\mathfrak{h}''$, satisfies that

$$C \cap \mathfrak{h}' = C \setminus \mathfrak{h}', \tag{B. 24}$$

Since $\forall\, t \in [0, 1]$, and $X, Y \in C$, $tX - (1 - t)Y \in C \cap \mathfrak{h}' = C \setminus \mathfrak{h}'' \subset C$, of where $tX - (1 - t)Y \in C$, $\forall\, t \in [0, 1]$. Then $B \cap C$ is connect. Then, if $X \in \Omega \cap C$, then exist $t > 0$ such that $tX \in B \cap C$. Thus $C \cap \Omega$, is connect (also $B \subset C \cap \mathfrak{h}$). ■

Theorem. B. 1. Be $\Omega = \{X + Y \,|\, X \in U, Y \in \Omega(\phi_1, \ldots, \phi_d, r)\}$. Let $T \in D'(\Omega')$, be such that $\dim I(\mathfrak{g}_C)T < \infty$, on Ω'. Then exist an analytic function $F_T = F$, on Ω', such that

1. $T = T_F$, on Ω',
2. If $\mathfrak{h}$, is a subalgebra of Cartan of $\mathfrak{g}$, then exist an analytic function β, on $\mathfrak{h}''$, which is an exponential polynomial on each connect component of $\mathfrak{h}''$, such that $F|_{\Omega \cap \mathfrak{h}'} = |D|^{-1/2}\beta$, more yet, if we extend F, to all Ω, doing correspond $F = 0$, on $\Omega - \Omega'$, then F, is locally integrable on Ω.

Proof. Assume $\mathfrak{h} = \mathfrak{h}_j$, then $\forall\, X \in \mathfrak{h}$, and $p \in S(\mathfrak{g})^G \cong I(\mathfrak{g})$

$$\psi_j^0(pT) = |D|^{-1/2}\underline{p}|D|^{-1/2}\psi_j^0(T), \qquad (B. 25)$$

Thus *dim* $\underline{I(\mathfrak{g})}(|D|^{-1/2}\psi_j^0(T)) < \infty$. But $S(\mathfrak{h}_C)$, is finitely generated like a $\underline{I(\mathfrak{g})}$-module. Then the lemma[13] implies that exist a function β_j, on $\Omega' \cap \mathfrak{h}_j$, whose restriction to each connect component is an exponential polynomial and is such that

$$\psi_j^0(T) = |D|^{-1/2}T_{\beta j}, \qquad (B. 26)$$

Given that

$$F|_{\Omega \cap \mathfrak{h}'}(X) = |D|^{-1/2}\beta(X) = F|_{\Omega j \cap \mathfrak{h}'j}(gH) = |D|^{-1/2}\beta_j(gH), \qquad (B. 27)$$

$\forall\ X \in \Omega'_j$, with $X = gH$, $H \in \mathfrak{h}'_j$, then $\beta(X) = \beta_j(gH)$. But $S(\mathfrak{h}_C)$, is finitely generated like a $\underline{I(\mathfrak{g})}$-module, thus

$$I(\mathfrak{g})\beta(X) = \underline{I(\mathfrak{g})}|D|^{-1/2}\beta_j(gH), \qquad (B. 28)$$

$\forall\ \underline{P} \in \underline{I(\mathfrak{g})}$. Thus $\beta_j(H) = \beta(X)$, in $\Omega' \cap \mathfrak{h}'_j$. Then if $F|_{\Omega \cap \mathfrak{h}'} = |D|^{-1/2}\beta$, then

$$\psi_j^0(T) = |D|^{-1/2}T_{\beta j} = \psi_j^0(T_F) = |D|^{-1/2}T_\beta,$$

Given that Thus $\beta_j(H) = \beta(X)$, in $\Omega' \cap \mathfrak{h}'_j$, then $T = T_{\Omega' \cap \mathfrak{h}'j} = T_{\Omega'} = T_F$, of where $T = T_F$, on Ω'.

Note that if we extend β, to Ω, by 0, then β, is locally bounded. Indeed, is sufficient only consider $\beta = 0$, on $\Omega - \Omega'$, of where

$$|D|^{-1/2}g|D|^{1/2}\beta_j(H) = 0,$$

which is equivalent to have that

$$|D|^{1/2}g|D|^{-1/2} = \beta_j(H),$$

Since *dim* $\underline{I(\mathfrak{g})}(|D|^{-1/2}\beta_j(T)) < \infty$, $\forall$ j, then exist α, such that $(\alpha = \alpha(0))$:

$$\alpha|D|^{1/2}g|D|^{-1/2} > \beta_j(H),$$

with $\alpha(0) = |\beta(X)|$, $\forall\ X \in \Omega_j' \forall$ j. Thus β, is locally bounded. Then by the corollary[14], $|D|^{-1/2}$, is locally integrable.

We define the open set

$$B^0 = \{X \in \Omega'_j \,|\, |\beta(X)| = \alpha(0) > 0\}, \qquad (B. 29)$$

[13] **Lemma:** Let U, be a connect open subset of $\mathbb{R}^n$. Let $T \in D'(U)$, be such that $S(\mathbb{R}^n)T$, is of finite dimension. Then exist $\mu_1, \ldots, \mu_p \in (\mathbb{R}^n)_C^*$, and $F \in \mathbf{C}[x_1, \ldots, x_n, e^{\mu 1}, \ldots, e^{\mu p}]$ such that $T = T_F$, on U. We have used a Lebesgue measure to $\omega_{\mathbb{R}}n$.

[14] **Corollary:** $|D|^{-1/2}$, is locally integrable on $\mathfrak{g}$.

that is to say, let be a neighborhood of the 0, such that $B^0 = \Omega \cap \mathfrak{h}_j'$. Since $F|_{\Omega'}$, is locally integrable then $F|_{B0}$, is locally integrable $\forall$ j. Then realizing an extension of β_j, by 0 in B^0, on Ω, and considering that $F = |D|^{1/2}\beta$, then F, is locally integrable in Ω. ■

The results that now we will give are extensions of the fundamental theorem of Harish-Chandra.

Let $X_1, \ldots, X_n$, be a base of $\mathfrak{g}$, and we define X^j, for

$$B(X_i, X^j) = \delta_{ij},$$

Let $\Box = \Sigma X_i X^j$. Then $\Box \in I(\mathfrak{g})$, ($\Box$ is a G-invariant differential operator). By definition

$$I(\mathfrak{g}) = \{D \in S(\mathfrak{g})^G | \, D = \Sigma X_i X_i, \; \forall \; \{X_j\}, \text{ a base of } \mathfrak{g}\}^{15}, \tag{B. 30}$$

since $S(\mathfrak{g})^G = S(\mathfrak{g}) \cap T(\mathfrak{g})$, where $T(\mathfrak{g})$, is a tensor algebra in $\mathfrak{g}$. Then $\Box = D_i{}^i \in T(\mathfrak{g})^0$, with $T(\mathfrak{g})^0 \subset T(\mathfrak{g})$. Thus $\Box \in I(\mathfrak{g})$.

Theorem. B. 2. Let Ω, be (B. 23). Let $T \in D'(\Omega)^G$, be such that

$$dim \; C[\Box]T < \infty,$$

on Ω. Let $F = F_T$, be like the given by theorem B. 1. Then $T = T_F$.

The demonstration of this result will take the reminder of the exposition. Before we have that to give some details of the demonstration, first we will develop some results on distributions on Ω, that will be required in $U \oplus \mathcal{N}$[16].

Note that if is a G-invariant polynomial on $\mathfrak{g}$, then $f(X) = f(X_s)$, $\forall \; X \in \mathfrak{g}$. Thus

$$\Omega \cap (z \oplus \mathcal{N}) = U \oplus \mathcal{N}, \tag{B. 31}$$

Indeed, by a side we know that

$$z(\mathfrak{g}) = \{X \in \mathfrak{g} | Ad(g)X = X, \; \forall \; g \in G\}, \tag{B. 32}$$

then $z \oplus \mathcal{N} = \{X \in \mathfrak{g} | Ad(g)X + ad^k X, \; \forall \; k \in \mathbb{Z}\}$. Since $U \subset \Omega \cap z(\mathfrak{g})$, then $\Omega \cap (z \oplus \mathcal{N}) \subset U \oplus \mathcal{N}$. For other side,

$$U \oplus \mathcal{N} = \{X + Y | X \in U \text{ and } Y \in \mathcal{N}\}, \tag{B. 33}$$

---

[15] This is a differential ideal and it is can extend or generalize to distributions on holomorphic vector bundles extending much of the results of classical differential ideals.
By $\mathfrak{u}$-cohomology and extending the concept of distribution on a complex holomorphic manifold subjacent in G/L, with L, a Levi subgroup in G, it is obtain a generalized $\bar{\partial}$-module.

[16] $\mathcal{N} = \{X \in \mathfrak{g} | \, I^+(\mathfrak{g}) = 0\}$.

Then $[X, Y] = 0$, $\forall\ Y \in \mathcal{N}$. But $Ad(g)X = X$, and $Y = ad^k X$, $\forall\ X \in \mathfrak{g}$, and $k \in \mathbb{Z}$. Since furthermore $[X, Y] = 0$, $\forall\ X \in \mathfrak{g}$. For other side Ω, is a convex space thus $\forall\ t \in \mathbb{R}^+$, $X + tY \in \Omega$. Thus $X + Y \in \Omega \cap (z \oplus \mathcal{N})$. Thus $U \oplus \mathcal{N} \subset \Omega \cap (z \oplus \mathcal{N})$. Then the equality among the sets $\Omega \cap (z \oplus \mathcal{N})$, and $U \oplus \mathcal{N}$, is followed.

We consider that $\mathfrak{g} = [\mathfrak{g}, \mathfrak{g}]$. Let $\mathcal{N} = O_1 \cup O_2 \cup \ldots \cup O_r$, be with $O_j = GX_j$, and O_1, open in $\mathcal{N}$, O_2, open in $\mathcal{N} - O_1$, etc. Let

$$\mathcal{N}_P = \bigcup_{I \geq P} O_1, \tag{B. 34}$$

Then $\mathcal{N}_P$, is closed in $\mathfrak{g}$. Let H, X, and Y, be a canonical base to a TDS algebra[17] $\mathfrak{u}$, in $\mathfrak{g}$, ($[H, X] = 2X$, $[H, Y] = -2Y$ and $[X, Y] = H$). Like $\mathfrak{u}$-module under ad, $\mathfrak{g}$, is the direct sum

$$\mathfrak{g} = \bigoplus_m V^m, \tag{B. 35}$$

such that $dim\ V^m = \mu_m + 1$, $\forall\ \mu_m \in \mathbb{Z}^+$. Then the eigenvalues λ, of the endomorphic equation

$$(adh - \lambda I_{vm}) = 0,\ \ \forall\ h \in V^m, \tag{B. 36}$$

are such that $Re\Lambda = \mu_m - 2k$, with $0 \leq k \leq \mu_m$. The proper μ_m–space is $\mathfrak{g}^Y \cap V^m$, and

$$XV^m = \Sigma \mathfrak{g}^Y \cap V^m, \tag{B. 37}$$

and $XV^j \cap XV^i = \varnothing$, $\forall\ i \neq j$. Then

$$\mathfrak{g} = \mathfrak{g}^Y \oplus [X, \mathfrak{g}], \tag{B. 38}$$

We consider $V = \mathfrak{g}^Y$. If $g \in G$, and if $Z \in V$, then we define $\Phi(g, Z) = g(X + Z)$, that is to say, the map defined by

$$\Phi : G \times V \rightarrow Ad(\Omega) \subset \Omega, \tag{B. 39}$$

with rule of correspondence

$$(g, Z) \mapsto Ad(g)(X + Y), \tag{B. 40}$$

Then the differential

$$d\Phi_{g,0}(g, V) = g(V + [X, \mathfrak{g}]) = \mathfrak{g}. \tag{B. 41}$$

Considering that $V = \mathfrak{g}^Y$, and considering the submersion

$$\psi: G \times (V \oplus [X, \mathfrak{g}]) \rightarrow \mathfrak{g}, \tag{B. 42}$$

with rule of correspondence

[17] Three-Dimensional-Split algebra.

$$(g, V + \underline{Y}) \mapsto Ad(g)X, \tag{B. 43}$$

we have in particular in a neighborhood of the $0 \in V$,

$$\psi = d\Phi_{g,0}(g, V), \tag{B. 44}$$

thus $Ad(g)X + Ad(g)adX^{\sim} \in \psi(G \times (V \oplus [X, \mathfrak{g}]))$. Then there is a subspace

$$V^{\sim} = \{X | X + V^{\sim} \subset \Omega\}, \tag{B. 45}$$

with Ω, the defined space explicitly as

$$\Omega = \{X + Y | X \in U \text{ (open), and } Y \in \Omega(\phi 1, \ldots, \phi 1, r)\},$$

In particular the images $Ad(g)X + Ad(g)adX^{\sim} \in X + V^{\sim} \subset \Omega$. Therefore

$$\Phi|_{G \times V\sim} = \psi(G \times (V \oplus [X, \mathfrak{g}])).$$

Then $\Phi|_{G \times V\sim} = d\Phi_{g,0}(g, V) = g(V + [X, \mathfrak{g}])$. Note $\Phi(G \times V^{\sim}) \cap \mathcal{N}_j$, is open in $\mathcal{N}_j$. Indeed, Φ, is suprajective in $G \times V^{\sim}$. Then $\Phi(G \times V^{\sim})$, is open in Ω. Thus $\forall$ j, $\Phi(G \times V^{\sim}) \cap \mathcal{N}_j$, is open in $\mathcal{N}_j$.

Let W[18], be a G-invariant subset of $\mathfrak{g}$, such that $W \cap \mathcal{N}_j = \mathcal{O}_j$. For definition of $V^{\sim}$, neighborhood of the 0, in V, we can re-define it on $\mathfrak{g}$, as

$$V^{\sim} = \{X \in V | d/dt \exp(tX)|_{t=0} = X(0)\},$$

But $X(0) \in d\psi_{g,0}(g, V) = \mathfrak{g}$. Also, $\Phi|_{G \times V\sim} = d\Phi_{g,0}(g, V)$, and given that $\Phi(g, V^{\sim}) \in W$, then V_j, is an open neighborhood of the 0, in $V^{\sim}$. Then $G \times V_j$, is open on $G \times V^{\sim}$. But Φ, is a submersion in $G \times V^{\sim}$, therefore exist an open set W such that

$$\Phi(G \times V_j) = W,$$

($\Leftrightarrow \forall$ j, Φ_j, is a submersion in $G \times V_j$), thus $\forall$ j, $\Phi(G \times V_j) \cap \mathcal{N}_j = \mathcal{O}_j$. If $X = 0$, then we consider $V_j = \Omega$. Then

3. Be $\mathcal{O}_j \subset \Omega$, and $X_j \in \mathcal{O}_j$. Be

$$V_j = \{X = X_j | \Phi(g, X_j) \in W, \forall g \in G\},$$

Then exist a neighborhood U_j, of 0, in V_j, such that if we consider $\Phi_j(g, Z) = g(X + Z)$, $\forall g \in G$, and $X \in V_j$, then

i. Φ_j, is a submersion in an open neighborhood Ω_j, of X in Ω,
ii. $\Omega_j \cap \mathcal{N}_j = \mathcal{O}_j$,
iii. $(X_j + U_j) \cap \mathcal{O}_j = \{X_j\}$.

---

[18] $W = \{X \in \mathfrak{g} | gX = X, \forall g \in G\}$

Anyone neighborhood of the 0, in V_j satisfies (i), and (ii). For which we demonstrate in V_j, that is satisfied (iii). If $X_j = 0$, we have $V_j = U_j$. Then we assume that $X = X_j \neq 0$. Let $\{X, Y, H\}$, be a base to a subalgebra TDS $\mathfrak{u}$, to X. Let

$$W^m = \bigoplus_{\Lambda <_\mu m} adV^m - \Lambda Ivm,$$

and let $W = \Sigma_{m\in N} W^m$. Then adX, is a linear isomorphism of W in $[X, \mathfrak{g}]$. Indeed, since

$$\mathfrak{g} = \bigoplus_m V^m,$$

then

$$[X, \mathfrak{g}] = \bigoplus_{m, \Lambda <_\mu m} [X, V^m] = [X, \bigoplus_m V^m] = [X, V] = [X, HW^m]$$

$$= [X, \mathfrak{g} \cap W^m] = [X, W^m]_{m\in N, \Lambda <_\mu m},$$

But this equality is satisfied if $\mathfrak{g} \cong_\phi W^m$, under the G-invariance on W^m, (that is to say, Ad(G), acting by 0 in W^m). Indeed, if we consider $Ad(G)W^m = W^m$, then the map

$$\varphi: G \times W^m \rightarrow U_j \text{ (open)}, \tag{B. 46}$$

with rule of correspondence

$$(g, X) \mid\!\rightarrow gX, \tag{B. 47}$$

is a linear isomorphism of $\mathfrak{g}$, in W^m, since exist W_0, neighborhood of the 0, in W and a neighborhood U′, of the 0, in U_j, such that

$$\Phi_j: W_0 \times U' \rightarrow W^{\sim}, \tag{B. 48}$$

with $W^{\sim} \subset W$, an open neighborhood of $X \in \mathfrak{g}$, and with rule of correspondence

$$(x, Z) \mid\!\rightarrow \Phi_j(\mathit{exp}x, Z), \tag{B. 49}$$

Let W_1, be a neighborhood of the 0, in W_0, such that $exp(adW_1)X$, is a neighborhood of X in $\mathcal{N}_j$. If we contract to the neighborhood W_0, in U′, we assume that

$$\Phi_j(\mathit{exp}W_0, U') \cap \mathcal{N}_j \subset \mathit{exp}(adW_1)X, \tag{B. 50}$$

Suppose that $Z \in U'$ and $X + Z \in O_j$. Then $X + Z \in O_j \cap \Phi_j(expW_0, U')$. Thus $X + Z = exp(adv)X$, $\forall v \in W_1$. Of where $\Phi_j(1, Z) = \Phi_j(expv, 0)$, then $expv = 1$. This implies that $v = Z = 0$. Then $(X_j + U_j) \cap O_j = \{X_j\}$. ■

Assume that $\mathfrak{g} = z \oplus [\mathfrak{g}, \mathfrak{g}]$. Let U_j, be like before. Consider also that $\Phi_j(g, U_j) = g(X + U_j)$. Indeed, we can to apply the before argument to the neighborhood by 0, in $\mathfrak{g}$, U_j, of U, and demonstrate that $\forall$ $Z \in U_j$, then $Z = 0$. But this is fulfilled trivially since $\mathfrak{g}^Y = z(\mathfrak{g})$. If $g \in G$, and

$Z \in U = z(\mathfrak{g})$, then $\forall\, X \in U_j$, $g(X + Y) \in U \oplus U_j$, where $U \oplus U_j$, is a neighborhood by 0, of $z(\mathfrak{g}) \oplus [\mathfrak{g}, \mathfrak{g}]$. Since $\Phi_j(g, U_j) \in \Phi(G \times U)$, is a diffeomorphism in 0, of $z(\mathfrak{g}) \times U'$, in a neighborhood of X, in $\mathfrak{g}$, then $\Phi_j(exp\, z(\mathfrak{g}), U')$, is an isomorphism. Let E, be a vector field on $\mathfrak{g}$, defined by

$$Ef(x + y) = (d/dt)(f(x + ty))|_{t=1}, \tag{B. 51}$$

$\forall\, x \in z(\mathfrak{g})$, and $y \in [\mathfrak{g}, \mathfrak{g}]$. If $x_1, \ldots, x_n$, are linear coordinates on $\mathfrak{g}$, such that $\{X_i\}_{i \leq q}$, are linear coordinates on $[\mathfrak{g}, \mathfrak{g}]$ and $\{X_i\}_{i > q}$, are coordinates on $z(\mathfrak{g})$, then

$$E = \Sigma_{i \leq q} x_i \partial/\partial x_i, \tag{B. 52}$$

If $x_i \in \{x_i\}$, then $\forall$ I, $x_i = \alpha_i + t\beta_I$, $\forall\, \alpha_i \in z(\mathfrak{g})$, and $\beta_i \in [\mathfrak{g}, \mathfrak{g}]$. Considering a restriction of $z \oplus [\mathfrak{g}, \mathfrak{g}]$, in B^0, (a neighborhood of the 0) then $\forall$ t = 0,

$$exp\,t[z(\mathfrak{g}) \oplus [\mathfrak{g}, \mathfrak{g}]]|_{B0} = exp(t\mathfrak{g})|_{t=0}, \tag{B. 53}$$

Then a canonical base of $E \in \mathfrak{g}$, in t = 0, is $\partial/\partial x_1, \ldots, \partial/\partial x_n$, and

$$E = \Sigma_{i > q} x_i + \Sigma_{i \leq q} x_i\, \partial/\partial x_i, \tag{B. 54}$$

But by (B. 53), $\forall\, E \in \mathfrak{g}$,

$$E = \Sigma_{i \leq q}(x_i + ty_i)|_{t=0}\ \partial/\partial x_i = E_i + tE_i \tag{B. 55}$$

Then $\Sigma_{i > q} x_i = 0$, of where $E = \Sigma_{i \leq q} x_i\, \partial/\partial x_i$.

Lemma B. 3. Let F, be a space of all the distributions with compact support $(z \oplus \mathcal{N}) \cap \Omega$. If $T \in F$, then $dim\mathbb{C}[E]T < \infty$, and the characteristic values of E, and F, are all real and strictly more little or lows that $-q/2$.

We fix a j, and let $O_j \subset \Omega$. Let $X \in z \oplus O_j$, be and let Φ_j, U_j, V, and Ω_j, be such that $\Phi_j(g, Z) = g(X + Z)$, $\forall\, g \in G$, and $X \in U_j$. Let

$$V_j = \{Z \in V^{\sim} |\ \Phi(g, Z) \in W,\ \forall\, g \in G\},$$

Let $V = z(\mathfrak{g})$. Assume that $O_j \neq \{0\}$. Let $y_1, \ldots, y_n$, be linear coordinates on $V \cap [\mathfrak{g}, \mathfrak{g}]$, such that $y_k(V \cap V^m) = 0$, if $m \neq k$. If $Z \in V$, then $Z = \Sigma_m Z_m$, with $Z_m \in V \cap V^m$. If we consider

$$adHZ_m + \mu_m Z_m = 0,$$

$\forall\, H = X + Z$, then

$$Z = \Sigma_m(1/2\mu_m + 1)Z_m, \qquad (1/2)H = X,$$

Indeed, only is necessary consider that $\forall\, H = X + Z$, $adHZ_m = -(\mu_m + \mu_m)Z_m$, of where

$$adHZ_m + (\mu_m + \mu_m)Z_m = 0,$$

of where $(1/2)Z = \Sigma_m(1/2\mu_m + 1)Z_m$. Then considering the map

$$\Phi_j : G \times (V \cap V^m) \rightarrow Ad(g)\Omega \subset \Omega, \quad (B.56)$$

with rule of correspondence

$$(g, Z) \mapsto g(X + Z), \quad (B.57)$$

whose differential is suprajective in $G \times (V \cap V^m)$, and $\forall$ j, we have that

$$(d\Phi_j)_{g, Z}((1/2)H, \Sigma_m(1/2\mu_m + 1)Z_m) = g(X + Z) = \Phi_j(g, Z), \quad (B.58)$$

$\forall$ $g \in G$, and $Z \in U_j$. Since Φ_j, is a submersion in $U_j \oplus \mathcal{N}_j$, we can to define distributions $T \in D'(\mathcal{N})$, and $E \in D'(U)$, such that

$$\psi^0(T) = E, \quad (B.59)$$

where E, is related with T. But $(d\Phi_j)_{g, Z}(G \times (V \cap V^m))$, implies that $\forall$ $Z \in \mathfrak{g}$,

$$\Phi_j^0(ET) = dg \otimes E = Z\Phi^0_j(T) = (\Sigma_m(1/2\mu_m + 1)y_m\partial/\partial y_m)\, \Phi_j^0(T), \quad (B.60)$$

The election of the system of neighborhoods of the 0, in $\mathfrak{g}$, given by the open sets in Ω, (sets foreseen in page 115) implies that if

$$\text{supp}\,(T) \subset (z \oplus \mathcal{N}_j) \cap \Omega,$$

Then

$$\text{supp}\,\Phi^0_j(T) \subset U \times \{0\},$$

Indeed, by definition

$$\text{supp}(T) = \{T \in F \mid T = X + Z,\ \forall\, X + Z \in (U_j \oplus \mathcal{N}_j) \cap \Omega\}, \quad (B.61)$$

and given that Φ_j, is a submersion of $G \times (V \cap V^m)$, in $Ad(g)\Omega$, in particular Φ^0_j, is a submersion such that

$$\Phi_j|_{U \times \{0\}}(T) = \Phi_j^0(ET) = Z\Phi^0_j(T) = ZE_j,$$

where

$$\text{supp}\Phi^0_j(T) = \{E \in F \mid E = \Sigma_{i \leq q} x_i\, \partial/\partial x_i,\ \forall\, x_i \in z(\mathfrak{g})\}, \quad (B.62)$$

Then $\text{supp}\Phi^0_j(T) \subset U \times \{0\}$.

Let

$$F_j = \{E^{\sim} \in F \mid \text{supp}\, E^{\sim} \subset (z \oplus \mathcal{N}_j) \cap \Omega\}, \quad (B.63)$$

We prove by descendent induction that is $T \in F_j$, then

$$dim C[E]T < \infty,$$

and the eigenvalues μ_m, of E, on F_j, are such that $\mu_m < -q/2$. Indeed, by the lemma[19], E, acts semi-simply (diagonalizable) on F_r, with strictly eigenvalues more little than $-q < -q/2$. We assume to j + 1, that this last affirmation is valid and we demonstrate this to F_j.

Let $T \in F_j$. Then $\Phi^0_j(T)$, have support in $U \times \{0\}$. Since

$$\Phi^0_j(ET) = (\Sigma_m(1/2\mu_m + 1)Z_m)\Phi^0_j(T), \qquad (B. 64)$$

with $Z_m = y_m\partial/\partial y_m$, then $\Phi^0_j(ET) = \Sigma_m(a_i + 1)E_j$, $\forall\ a_i \in \mathbb{R}^+$ (I = 1, 2, …, s) then $\Phi^0_j(ET)$, acts semisimply on the space of tempered distributions $D'(U) = F'_j \subset F_j$, that is to say;

$$\Phi^0_j(\Pi(E - a_i)T) = 0,$$

We call $\Sigma_m(\mu_m + 1) = q$. Such $\Sigma\mu_m = q - d$. Thus $-a_i \geq ½(d + q)$. Thus of the implication

$$\text{If } supp(T) \subset (z \oplus \mathcal{N}_j) \cap \Omega, \text{ then } supp\Phi^0_j(T) \subset U \times \{0\},$$

we have that if $suppE^{\sim} = suppET \subset F_j$, then

$$supp(\Pi(E - a_i))T \subset F_{j+1},$$

Thus the property is established $\forall$ j. Then $\forall$ j,

$$dim\mathbb{C}[E]T < \infty,$$

and all the eigenvalues μ_m, of E, on F, are real and more strictly little that $-q/2$.

Let $\omega(X + Z) = B(X, X)$, be $\forall\ Z \in z$, and $X \in [\mathfrak{g}, \mathfrak{g}]$. Let X_i, be a base of $\mathfrak{g}$, such that $x_i \in z(\mathfrak{g})\ \forall\ i > q$, and $B(xi, xj) = \varepsilon_i\delta_{ij}$, with $\varepsilon_i = \pm 1$.

Let $\square_1 = \Sigma_{i \leq q}\varepsilon_i\partial^2/\partial x_i^2$, and $\square_0 = \square - \square_1$. We see to ω, like a differential operator under multiplication. Be $h = E + (q/2)I$, $x = -1/2\omega$, and $y = \square_1$. Then a direct calculus gives

$$[h, x] = 2x,\ [h, y] = -2y,\ [x, y] = h,$$

Lemma. B. 4. If $T \in F$, and if p, is not a null polynomial in a variable then $p(\square_1)T = 0$, implies that $T = 0$.

Proof. Indeeed, consider the commutative ring

---

[19] **Lemma.** Let a_j, be real numbers non-negative to j = 1, …, n. Let $D = \Sigma(a_j + 1)E_j$. Then D act semi-simply on the spaces $D'_{U1}(U)$, with real eigenvalues $-\lambda$, such that $\lambda \geq n + \Sigma aj$.
D'U1(U), is the space of distributions on U supported on $U1 \times 0$.

$$\mathbb{C}[H] = \{p \in P(\mathfrak{g})^G \mid p(D) = p\Sigma_{i \leq q} x_i \partial/\partial x_i,\ \forall\ x_i \in Z(\mathfrak{g})\}, \tag{B. 65}$$

and consider the torsion element T∈F, where T, is a $\mathfrak{u}$-module that satisfy the hypothesis of the corollary[20] then $\mathbb{C}[H]T = 0$, ∀ T∈F. In particular to someone $\square_1 \in DO(\Omega') \cong I(\mathfrak{g})$,

$$p(\square_1)T = 0,$$

then T = 0, to someone $p(\square_1) \in \mathbb{C}[H]$, which is equivalent to that

$$Tor(F_j, F_{j+1}) = 0. \tag{B. 66}$$

■

Now we demonstrate the following lemma that is useful in the final of the demonstration of the regularity theorem.

Lemma B. 5. If S∈F, and if p, is a polynomial not null in a real variable such that $p(\square)S = 0$, then S = 0.

Proof. First we demonstrate that if S∈F, $\zeta \in \mathbb{C}$, if $(\square - \zeta)S = 0$, then S = 0. Let $S = \Sigma S_\mu$, be with $(h - \mu)^d S_\mu = 0$, to someone $d \in \mathbb{Z}^+$. Then

$$0 = (\square - \zeta)S = \Sigma(\square_0 - \zeta)S_\mu + \Sigma\square_1 S_\mu,$$

Let λ, be the minimal μ, such that $S_\mu \neq 0$, that is to say; the minimal eigenvalue of $S_\mu \neq 0$. Since

$$(h - (\mu - 2))^{d'} \square_1 S_\mu = 0, \tag{B. 67}$$

∀ h∈DO(Ω'), and (μ – 2), the corresponding value of S_μ, ∀ $d' \in \mathbb{Z}^+$, with d' < d, then

$$\square_1 S_\lambda = 0,$$

By the corollary of the foot page 10, implies that $S_\lambda = 0$. But this contradicts the fact of that $S_\mu \neq 0$, to someone μ, minimal. To demonstrate the lemma B. 5, apply the hypothesis of induction to the grade of p.

If $grp = 0$, then the result is trivial. Suppose that the result to p ≠ 0, such that $grp = d - 1 \geq 0$, and we demonstrate the validity of the affirmation of the lemma to $grp = d$, ∀ $d \in \mathbb{Z}^+$. Indeed, if $grp = d$, then $p(t) = (t - \zeta)q(t)$, ∀ $\zeta \in \mathbb{C}$, and q(t), a polynomial of grade d – 1. But if $q(\square)S = 0$, then S = 0, for hypothesis of induction. Then

$$0 = p(\square)S = (\square - \zeta)(q(\square)S).$$

But this affirm that $q(\square)S = 0$, then S = 0. ■

[20] **Corollary.** Let M, be $\mathfrak{u}$-module such that if m∈M, then dim $\mathbb{C}[H]m < \infty$, and such that the eigenvalues of H, on M are real and strictly minors than 0. Then the action of $\mathbb{C}[Y]$, is a free torsion on M.

Note: T is a element of torsion since $(H - q)T = 0$, for someone $H - qI \in \mathbb{C}[H]$, $(H - q) \neq 0$.

APPENDIX C: Some Elements of the Asymptotic Behavior of the Matrix Coefficients

In this appendix we treat of highlight the analytic part of the real reductive group in the endomorphism algebra whose coefficients are the maximum weight of the discrete principal series and that are the coefficients of the cohomology of the exact succession of the ($\mathfrak{g}$, K)-modules that are finitely generated like U($\mathfrak{n}$)-modules, which induces the Hilbert representation of G, of finite dimension useful to the representation of a Lie group of infinite dimension.

Let G, be a real reductive group, and we will do the identification $G^0 = {}^0(G^0)$, along of this appendix, where G^0, is the identity component of G, (*remember that* ${}^0G = \{g \in G | \chi^2 = 1, \forall \chi \in X(G)\}$. ${}^0G = {}^0ANK$, G^0, *is the identity component of* G, *and* X(G), *is the space of continuous homomorphism of* G, *in the multiplicative group* $R^* = (R, \bullet)/\{\}$), then ${}^0(G^0)$, is the identity component of ${}^0G$. Then if

$$G^0 = \{g \in G | Ad(g) = I, \forall\ Ad \in End(G)\}, \qquad (C.\ 1)$$

In particular

$${}^0(G^0) = \{g \in G | \chi(g) = 1, \forall\ \chi \in X({}^0G)\}, \qquad (C.\ 2)$$

Let Δ_0, be the set of semi-simple roots of $\Phi(P, A)$. Let F, be a subset of Δ_0, and be (P_F, A_F), the corresponding canonical parabolic pair, that is to say; explicitly

$$\Phi(P_F, A_F) = \{(p, a) \in M_F | M_F = A_F \times {}^0M_F, \text{ and } P_F = M_F N_F\}, \qquad (C.\ 3)$$

Lemma. C. 1. Be $V \in \mathcal{H}$. Then the module $V/\mathfrak{n}_F V$, is an admissible module finitely generated like a $(\mathfrak{m}_F, P_F \cap K)$-module.

Proof. See [28], and [35].

Lemma. C. 2. Let $V_1^{\sim} \subset V_2^{\sim} \subset \ldots$, be an increasing chain of sub-modules of $V^{\sim}$.

Proof. Let $V_j = \{v \in V | V_j^{\sim}(v) = 0\}$. Then $V_1 \supset V_2 \supset \ldots$, is a decreasing chain of submodules of V. Indeed, the exact succession in $\mathcal{H}$,

$$0 \to V_j \to V \to V/V_j \to 0, \quad \forall\ j > k, \qquad (C.\ 4)$$

induces an exact succession in $\mathcal{V}$, given for

$$0 \leftarrow V_j^{\sim} \leftarrow V^{\sim} \leftarrow V^{\sim}/V_j^{\sim} \leftarrow 0, \quad \forall \ j > k, \tag{C. 5}$$

which $\forall \ j \in \mathbb{Z}^+$, $\ker T_j \supset \mathfrak{n}^j V^{\sim}$. By the theory of Jacquet modules, V is a $U(\mathfrak{n})$- module, finitely generated like a $\mathfrak{g}$-module whose finite length is the of $(\mathfrak{g}, K)$-module. Then exist $k \in \mathbb{Z}^+$, minimal such that $V_j = V_k$, $\forall \ j > k$. But for the third application of the Jacquet modules, having a $V^{\sim}$, like a Jacquet module of V, V, is admissible and the Jacquet module of V, is

$$j(V) = \{\mu \in V^{\sim} \mid \mu(V_j) = 0\}, \tag{C. 6}$$

thus $V_j^{\sim} = V_k^{\sim}$, $\forall \ j > k$. Then $V^{\sim}$, is finitely generated like a $U(\mathfrak{n})$-module, then by a theorem that affirm that a $(\mathfrak{g}, K)$-module that is finitely generated like $U(\mathfrak{n})$-module is admissible, $V^{\sim}$, is admissible ($V^{\sim} = \oplus V(\gamma)^*$). Then $V^{\sim}$, is a $(\mathfrak{g}, K)$-module finitely generated admissible, thus $V^{\sim} \in \mathcal{H}$. ■

Let (π, H), be a Hilbert representation of G. Let $(H^{\infty})' = \mathcal{L}(H^{\infty}. \mathbb{C})$. If $g \in G$ (respectively $X \in \mathfrak{g}$), and $\mu \in (H^{\infty})'$, then we define the map

$$G \times (H^{\infty})' \rightarrow H^{\infty}, \tag{C. 7}$$

whose rule of correspondence is

$$(g, \mu) \mapsto g\mu, \tag{C. 8}$$

where $\forall \ v \in (H^{\infty})'$; $g\mu(v) = \mu(\pi(g^{-1})v)$, (respectively $X\mu(v) = -\mu(\pi(X)v$, $\forall \ X \in \mathfrak{g}$), then for the elemental theory of representations, is had that

$$gX\mu = (Ad(g)X)g\mu, \tag{C. 9}$$

$\forall \ \mu \in (H^{\infty})'$, $X \in \mathfrak{g}$, and $g \in G$. Indeed, consider the map

$$\pi : U^j(\mathfrak{g}) \otimes (H^{\infty})' \rightarrow H^{\infty}, \tag{C. 10}$$

whose rule of correspondence is

$$g \otimes v \mapsto \pi(g)v, \tag{C. 11}$$

$\forall \ v \in (H^{\infty})'$. Then

$$gXv = \pi(g)\pi(X)v, \tag{C. 12}$$

but for be $(H^{\infty})'$, a subspace finitely generated like a $(\mathfrak{g}, K)$-module of H^{∞}, then

$$\pi(g)\pi(X)v = \pi(Ad(g)X)\pi(g)v, \tag{C. 13}$$

thus, if $v \in (H^{\infty})'$, $X \in \mathfrak{g}$, and if H, is the generated for $\pi(K)v$, then

$$\pi(Ad(g)X)\pi(g)v = \pi(Ad(g)X)gv,$$

$\forall\ \mu\in(H^{\infty})'$, $X\in\mathfrak{g}$, and $g\in G$.

Let $(H^{\infty})'_K = \{\mu\in(H^{\infty})' \mid K\mu$, generate a subspace of finite dimension$\}$, then the before identity imply that $(H^{\infty})'_K$, is a $(\mathfrak{g}, K)$-module. Indeed, $\forall\ k\in K$, $\mu\in(H^{\infty})'$, $X\in\mathfrak{g}$

$$kX\mu = (Ad(k)X)k\mu,$$

Then the elements $\pi(K)v = Kv$, generate a subspace of finite dimension of $(H^{\infty})'$, whose actions $K\mu$, are continuous. In particular, Kv, are continuous and generate a all subspace of the module $(H^{\infty})'_K$. If $Y\in\mathfrak{t}$, and $v\in(H^{\infty})'$, then

$$(d/dt)\exp(tY)v|_{t=0} = Yv,$$

thus $(H^{\infty})'_K$, is a $(\mathfrak{g}, K)$-module. Let be the map

$$\sigma : H \to H', \tag{C. 14}$$

whose rule of correspondence is

$$v \mapsto \sigma(v), \tag{C. 15}$$

Then $\forall\ w\in H$, $\sigma(v)(w) = \langle v, w\rangle$. Then σ, is a linear continuous and conjugate isomorphism of H, in H'. H', is the space of the linear functional on H, that is to say; $H' = \mathcal{L}(H, \mathbb{C})$, is endowed like Banach space of an inner product that is continuous in all point of H. Then the map

$$H \times H' \to H', \tag{C. 16}$$

whose rule of correspondence

$$(v, w) \mapsto \langle v, w\rangle, \tag{C. 17}$$

is a linear bijection such that ${}^{c}\sigma(v) = {}^{c}\langle v, w\rangle = \langle v, w\rangle = \sigma(w)$. Thus the map σ, is a conjugate continuous linear isomorphism of H in H'.

Lemma. C. 3. If (π, H), is admissible then $(H^{\infty})'_K = \sigma(v)_K$. Furthermore, $(H^{\infty})'_K = (H_K)^{\sim}$.

Proof. If (π, H), is admissible then $\forall\ \gamma\in K^{\wedge}$, $\dim H(\gamma) < \infty$, then

$$\sum_{\gamma\in K^{\wedge}} H(\gamma)v = (H^{\infty})'_K, \tag{C. 18}$$

If $w\in H$, then

$$\sum_{\gamma\in K^{\wedge}} H(\gamma)v = H\otimes H_K = \sigma(H_K), \tag{C. 19}$$

since to all $\forall$ $w \in H$, and $v \in H_K$, $\sigma(v) \in \sigma(H_K)$. Then for (B. 18), and (B. 19), it is have that

$$(H^{\infty})'_K = \sigma(H_K), \tag{C. 20}$$

Then $\sigma(H_K) = (H_K)^{\sim}$. Given that

$$\sigma(H_K) = \Sigma_{\gamma \in K^{\wedge}} H(\gamma)v = \oplus_{\gamma \in K^{\wedge}} H^*(\gamma) = H^*, \tag{C. 21}$$

Since H^*, is a K-module, and for definition

$$(H_K)^{\sim} = \{v \in H^* | \ Kv, \text{ generates a subspace of finite dimension}\}, \tag{C. 22}$$

Using the map

$$K \times H^* \to H^{\sim}, \tag{C. 23}$$

whose rule of correspondence is

$$(k, v) \mapsto kv, \tag{C. 24}$$

$\forall$ $v \in H^*$, then $\sigma(H_K) = (H_K)^{\sim}$. Then $(H^{\infty})'_K = (H_K)^{\sim}$. ∎

But this is posible given that ($\mathfrak{m}_F$, P_F)-modules of $V/\mathfrak{n}_F V$, are admissible like U($^*\mathfrak{n}_F$)-modules and $\forall$ $V^{\sim} \in H$, with

$$V^{\sim} = \{\mu \in V^* \,|\, \Sigma_j K\mu_j = w, \text{ with } w \in W, \text{ and } W \subset V^*\}, \tag{C. 25}$$

we have that $V^{\sim} = (H_K)^{\sim} = (H^{\infty})'_K$ y $(H^{\infty})'_K = \sigma(H)_K$, with

$$\sigma \in \Lambda(H, H') = isom(H, H'),$$

and $(H^{\infty})'_K$, is a M-module. Then considering the inequalities, using images in $\sigma(H^{\infty})'_K$, to know,

$$|(\mu(\pi(a))v)| \leq a^{\delta}\sigma'_{\mu}(v) \ \forall \ v \in H^{\infty}, \text{ and } a \in CL(A^+),$$

and

$$|(\mu(\pi(a))v)| \leq a^{\delta - \alpha}\gamma'_{\mu}(v) \ \forall \ v \in H^{\infty} \text{ and } a \in CL(A^+),$$

we can estimate the terms of the integral:

$$F(t, a'; v) = \exp(-tB)F(t_0, a'; v) - \exp(-tB)\int_0^t \exp(sB)\, G(s, a'; v)ds, \tag{C. 26}$$

that of the inequalities, we obtain

$$\|F(t, a'; v)\| \leq (a')^{\delta}\beta(v), \ \forall \ a' \in CL(A^+), \tag{C. 27}$$

with β, a continuous seminorm on H^{∞}. Then we obtain

$$||F(t, a'; v)|| \leq \exp|(\delta(H) - 1)t|\ (a')^{\delta}\beta'(v),\ \forall\ a' \in CL(A^{+}), \tag{C. 28}$$

with β', a continuous seminorm on H^{∞}. But

$$||\exp(sB)|| \leq C(1 + |\sigma|)^{p} e^{sRez}\ \ \forall\ \ s \in \mathbb{R},\ p \leq d,$$

which is immediate of $(B - zI)^{P} = 0$, to $B = [b_{nk}]$. Such calculation implies that

$$||F(t, a'; v)|| \leq C(1+t)^{P} e^{-tRez} (a')^{\delta} \beta(v)(1 + \int_{0}^{t} e^{s(Rez + d(X) - 1)} ds), \tag{C. 29}$$

to some continuous seminorm β, on H^{∞}, y $C > 0$. Observe that $(1 + s)^{p} e^{-\varepsilon s}$, is bounded by $C > 0$, to $\varepsilon > 0$, and $s \geq 0$. Thus also

$$||F(t, a'; v)|| \leq C(1 + t)^{p} e^{-tRez} (a')^{\delta}\beta(v) + C(1 + t)^{p} e^{s(Rez + \delta(X))} (a')^{\delta}\beta(v), \tag{C. 30}$$

$\forall\ t > 0$, β, a continuous seminorm on H^{∞}, and $C > 0$.

The cases exist:

Case I. If $\delta(X) - 2/3 \leq \Lambda(X)$, then exist a continuous seminorm β, on H^{∞}, such that

$$||F(t, a'; v)|| \leq C(1 + t)^{p} e^{t\Lambda(X)} (a')^{\delta}\beta(v)\ \ \forall\ \ t \geq 0,$$

Case II. If $\delta(X) - 2/3 > \Lambda(X)$, then in (I), we replace δ, by $\delta - (1/2)\alpha$, and we iterate the steps to the inequality (III). Of this way, we reduce the case II to case I.

How can refines this technique to demonstrate the asymptotic behavior of the developments of the matrix coefficients $(\mu(\pi(a))v)$, of the finite generated Hilbert representation and admissible required to induce representations?

Let $F \subset \Delta_{0}$, be with Δ_{0}, a roots space in the system of roots $\Phi(P_{F}, A_{F})$ corresponding to canonical parabolic pair (P_{F}, A_{F}).

Then the asymptotic behavior of the matrix coefficients of the induced representations to operators of infinite dimension is established while that $\sigma(\pi(\exp(tX)v)$, be asymptotic, to know

$$\Sigma_{\mu \in E0} \exp(t\mu(X)) \Sigma_{Q \in C+} \exp(-tQ(X)) P_{\mu, Q}(TX, \sigma; v),$$

when $t \to +\infty$.

finally, given that M, is a connect group on all element of G, the coefficients of the representations of the operators of infinite dimension in M, have that satisfy also this bounding condition since G is a covering of M

APPENDIX D: Coexeter Diagrams to some Integral Transforms in Representation Theory

We realize a little digression of the irreducible roots systems.

Def. D. 1. Φ, is irreducible if cannot be partitioned in the union of proper subsets such that each one of their roots in each set be orthogonal to each root of the other.

Examples: The systems of roots B_2, and G_2:

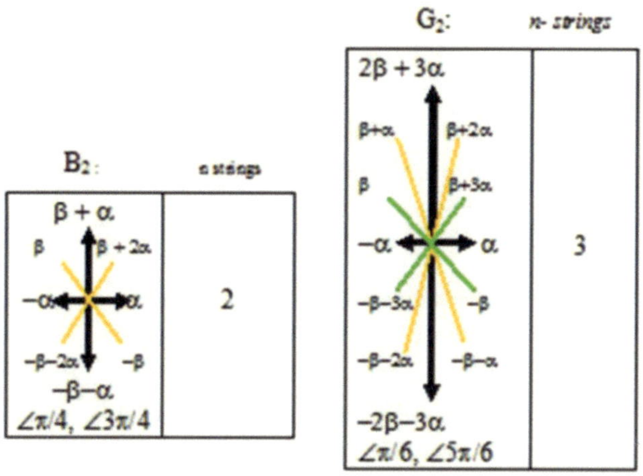

a) b)

Figure 1. Root systems to two and three strings Φ (= B_2, G_2).

One that is not irreducible is $A_1 \times A_2$.

Let Δ, a base of Φ. Φ, will be irreducible if only yes Δ, cannot be partitioned as it has been mentioned earlier. We demonstrate the implication:

⇒). If Δ, cannot be partitioned then Φ, is irreducible.

Proof. We demonstrate the contraposition: If ~Q ⇒ ~P. We suppose that Φ, admits a partition in subsets or classes

$$\Phi_1 \cup \Phi_2 = \Phi, \qquad \text{(D. 1)}$$

With $(\Phi_1, \Phi_2) = 0$ (is to say, orthogonal between they), to be not that Δ, can entirely to be included in Φ_1, or in Φ_2, which induce a similar partition of Δ, but $\Delta \subset \Phi_1$, then $(\Delta, \Phi_1) = 0$, or $(\Delta, \Phi_2) = 0$, since Δ, develop or generate to E. This demonstrates that Δ, is partitioned $(\Delta = \Delta_1 \cup \Delta_2)$. ■

$\Rightarrow$). If Φ, is irreducible then Δ, is not partitioned.

Proof. Let Φ, be irreducible, but $\Delta = \Delta_1 \cup \Delta_2$, with $(\Delta_1, \Delta_2) = 0$. But by the theorem on the action of the Weyl group on Φ, is had that, if $\alpha \in \Phi$, $\sigma \in W$, then $\sigma(\alpha) \in \Delta$, that is to say, each root is conjugated to a simple root such that $\Phi = \Phi_1 \cup \Phi_2$, with Φ_i, the set of roots, having each ith-component a conjugated in Δ_1. Remembering that $(\alpha, \beta) = 0$, $\forall \ \alpha \in \Phi$, and $\beta \in \Delta$, then $\sigma_\alpha \sigma_\beta = \sigma_\beta \sigma_\alpha$ (since $\sigma_\alpha = \sigma_\beta \sigma_\alpha \sigma_\beta^{-1}$). Thus the Weyl group W, is generated by the reflections σ_α $(\alpha \in \Delta)$. But the formula to the reflections affirms clearly that each root in Φ_i, is given of one in Δ_i, for addition or subtraction of elements of Δ_i. Thus, said element fall in the subspace E_i, of E, generated by Δ_i, and we see that $(\Phi_1, \Phi_2) = 0$. This implies that $\Phi_1 = \varnothing$, or $\Phi_2 = \varnothing$, of where $\Delta_1 = \varnothing$, or $\Delta_2 = \varnothing$. Thus $\Delta \neq \Delta_1 \cup \Delta_2$, with $\Delta_i = \varnothing$, with i = 1, 2. Thus Δ, it does not admit to be partitioned. ■

Lemma D. 1. Let Φ, be irreducible. Relative to partial order $\prec$, exist a unique maximal root (that be maximal means that is the root of maximal height in Δ), β, (in particular $\alpha \neq \beta$, then $ht\alpha < ht\beta$, and $(\beta, \alpha) \geq 0$, $\forall \ \alpha \in \Delta$). If $\beta = \Sigma K_\alpha \alpha$, $\forall \ \alpha \in \Delta$, then all the $K_\alpha \geq 0$.

Proof. Let $\beta = \Sigma K_\alpha \alpha$, $\forall \ \alpha \in \Delta$, maximal with the relation of order $\succ$, evidently $\beta \succ 0$. If $\Delta_1 = \{\alpha \in \Delta \mid K_\alpha > 0\}$, and $\Delta_2 = \{\alpha \in \Delta \mid K_\alpha > 0\}$, then $\Delta = \Delta_1 \cup \Delta_2$, is partition. Suppose that Δ_2, is not vanish, that is to say $\Delta_2 \neq \varnothing$. Then $(\alpha, \beta) \leq 0$, $\forall \ \alpha \in \Delta_2$, (by lemma of simple roots) of where Φ, is irreducible then at least $\alpha \in \Delta_2$, can be orthogonal to Δ_1, of where to some α', arbitrary of Δ_1, $(\alpha, \alpha') < 0$, of where $(\alpha, \beta) < 0$. This implies by the lemma that give criteria through of the functional (α, β), on the character of the roots $\alpha - \beta$, and $\alpha + \beta$, is a root, but clearly $\alpha + \beta > \beta$, which contradict the maximalist of β. Thus Δ_2, cannot have elements, that is to say, $\Delta_2 \neq \varnothing$, of where all $K_\alpha > 0$. This demonstrate also that $(\alpha, \beta) \geq 0$, $\forall \ \alpha \in \Delta$ (with $(\alpha, \beta) > 0$, to some α, where Δ, generates to E).

Now we demonstrate that in the system $(\Phi, \prec)$, exist a unique maximal root β. Consider other root with the mentioned property in the system of irreducible roots with partial order $(\Phi, \prec)$. Let such root β'. The precedent argument (of the before demonstration) we apply it to β', of the form that wrap at least a root $\alpha \in \Delta$, (with positive coefficients) to which $(\alpha, \beta) > 0$. It follows that $(\beta', \beta) > 0$, and $\beta - \beta'$, is a root by the lemma on the criteria of the functional (α, β), to elemental combinations and operations of roots. Thus $\beta \neq \beta'$. But if $\beta - \beta'$, is a root then or $\beta \prec \beta'$, or $\beta \succ \beta'$, which result be a contradiction, since β', is maximal. Then $\beta = \beta'$, of where β, is unique. ■

Lemma D. 2. Let Φ, be irreducible. Then W, act irreducible on E. In particular, the W orbit[21] of a root α, generates E.

Proof. The development of a W-orbit of a root is a W-invariant subspace of E (non-vanishing) such that the W-orbit of a root generates to E. Let $E' \subset E$, and W-invariant. Then there is a subspace E^* ($E^* \oplus E' = E$) W-invariant. Likewise, if $\alpha \in \Phi$, such that $\alpha \in E'$ and $E' \subset \mathfrak{P}_{\alpha}$, then $\sigma_{\alpha}(E') = E'$, therefore $\alpha \notin E'$, then $\alpha \in E^*$, such that every root is found inside of a subspace or of the other. This divides to Φ, in orthogonal subsets, forcing them to that one or other let be vanishing. Of the affirmation of that Φ, generates to E, is concluded then that $E' = E$. Thus $\sigma(E') \subset E'$, $\forall\ \sigma \in W$, and W acts irreducible on E. ■

Notes: In other words the subspace of the reflexions of the subspace E′, is an irreducible root system and therefore all automorphism $\sigma \in W$, acts irreducible[22]

Lemma D. 3. Let Φ, be irreducible. Then at least two roots of different length happen in Φ, and all the roots of length given are conjugated under W.

Proof. Let α, β, be roots of different length, then not all σ(α), $\forall\ \sigma \in W$, can be such that $(\sigma(\alpha), \beta) = 0$, where σ(α), generates to space E (Lemma D. 2). If $(\alpha, \beta) \neq 0$, (where by the lemma on root spaces theory, that says: *If* Δ, *is a base of* Φ, *then* $(\alpha, \beta) \leq 0$, $\forall\ \alpha \neq \beta$, *in* Δ, *and* $\alpha - \beta$, *is not a root*, we know that the possible radius of the square of the root length of α, and β, are 1, 2, 3, ½, 1/3[23]).

Now, if α, and β, have equal length to replace a of these for their W-conjugated we can assume that the roots α, ${}^{c}\alpha = \gamma\sigma(\alpha)\gamma^{-1}$, $\forall\ \gamma \in \underline{C(\Delta)}$, and $\sigma(\alpha) = {}^{c}\alpha \in W(E, \Delta)$, are different and non-orthogonal. By the same lemma mentioned in this demonstration is deduced that

$$\langle\alpha, \beta\rangle\langle\beta, \alpha\rangle = \pm 1. \qquad \text{(D. 1)}$$

■

Lemma D. 4. Let Φ, be irreducible, with two different roots. Then the maximal root β, of the Lemma D. 1, is the major length.

---

[21] Remember that the orbit as topological concept of a class space is a lateral class, where every element of W, is the class of an element in N(T), such that $\sigma w\ t = \sigma\ t\ \sigma^{-1}$.

[22] This is equivalent to say also that W, acts canonically on GL(*n*, E)

[23] Only remember that from $(\alpha, \beta) = \|\alpha\|\ \|\beta\|\cos\theta$, had been deduced:

$$\langle\beta, \alpha\rangle = 2(\beta, \alpha)/(\alpha, \alpha) = 2\{\|\beta\|/\|\alpha\|\}\cos\theta,$$

then $(\alpha, \alpha)\langle\beta, \alpha\rangle/2 = (\beta, \alpha)$, where

$$\{\|\alpha\|\ \|\beta\|\|\alpha\|\ \cos\theta\}/2 = (\beta, \alpha).$$

Proof. Indeed, let $\alpha\in\Phi$, be arbitrary. Is enough demonstrate that the criteria of comparison that takes the computing of the functional $(\alpha, \beta) \neq 0$, $\forall\ \alpha, \beta\in\Phi$, that $(\beta, \beta) \geq (\alpha, \alpha)$, that is to say, $\|\beta\| > \|\alpha\|$, $\forall\ \ \beta\in\Phi$. Then we substitute α, for a W-conjugated root that is inside the closure of the fundamental Weyl camera (relative to Δ).

For other side, we know by the lemma D. 2, that $\beta - \alpha > 0$. Thus to any γ–regular, that is to say, $\gamma\in\underline{C(\Delta)}$,[24] $(\gamma, \beta - \alpha) \geq 0$. This fact applied to the cases $\gamma = \beta$, and $\gamma = \alpha$, is had that

$$(\gamma, \gamma - \alpha) = (\beta, \beta) \geq (\gamma, \gamma - \beta) = (\alpha, \alpha), \tag{D. 2}$$

Therefore $(\beta, \beta) \geq (\beta, \alpha) \geq (\alpha, \alpha)$, $\forall\ \ (\beta, \alpha) \geq 0$, and $\alpha, \beta\in\Phi$. ■

We pass now to Classification study of the root systems. Said classification will serve after as a essential base in the Lie groups classification of arbitrary dimension.

Let Φ, a root system of range l, W, their Weyl group and Δ, a base of Φ.

Def. D. 2. (Cartan matrix of Φ). Fixing a arrangement $(\alpha_1, \ldots, \alpha_l)$ of simple roots. The matix $(<\alpha_I, \alpha_j>)$, is called a Cartan matrix of Φ. Their enters are called Cartan integers.

To systems of 2-range, we have the following matrices:

$$A_1 \times A_2 = \begin{pmatrix} 2 & 0 \\ 0 & 2 \end{pmatrix}, \tag{D. 3}$$

$$A_2 = \begin{pmatrix} 2 & -1 \\ -1 & 2 \end{pmatrix}, \tag{D. 4}$$

$$B_2 = \begin{pmatrix} 2 & -2 \\ -1 & 2 \end{pmatrix}, \tag{D. 5}$$

$$G_2 = \begin{pmatrix} 2 & -1 \\ -3 & 2 \end{pmatrix}, \tag{D. 6}$$

Mapping Φ, in Φ', and satisfying that $<\phi(\alpha), \phi(\beta)> = <\alpha, \beta>$ $\forall\ \ \alpha, \beta\in\Phi$, the Cartan matrix of Φ, determines to Φ, by the isomorphism ϕ.

---

[24] Remember that the Weyl cameras in a sense more restricted, is all relative camera to Δ. If further more is delimited for two hyper-planes then this is a closure of the Weyl camera in the more general sense. This is called a fundamental domain or fundamental Weyl camera. All fundamental Weyl camera can extending to a open connect region of the space E. Every point of $\underline{C(\Delta)}$, is W-conjugated to a point in E, to know

$$\forall\ \ \gamma\in\underline{C(\Delta)}, \text{ and } \sigma\in W\gamma^{-1}\sigma\gamma\in\underline{C(\Delta)},$$

that is to say, $\gamma^{-1}W\gamma\in\underline{C(\Delta)}$.

The shape of the matrix will depend of the elected arrangement, although this is not relevant. Important is that the Cartan matrix is independent of the election of Δ; this for the part b), of the theorem on the permutation of Weyl cameras or bases of Φ, then if Δ′, is other base of Φ, then σ(Δ′) = Δ, ∀ σ∈W (such that W, act transitively on bases).

Proposition D. 1. Let $\Phi' \subset E'$, be other root system with base $\Delta' = \{\alpha'_1, \ldots, \alpha'_l\}$. If $<\alpha'_i, \alpha'_j> = <\alpha_i, \alpha_j>$, to $1 \leq i, j \leq l$. then the bijection $\alpha_i \mapsto \alpha'_I$, is extended univocally to the isomorphism $\phi: E \rightarrow E'$.

Proof. Using the extension of the Coxeter graphs and the properties of correspondence between bases and Weyl cameras is demonstrated the result. ■

If α, β∈Φ, α ≠ β, then is acquaintance that $<\alpha, \beta><\beta, \alpha> = 0, 1, 2$ 0r 3, according to the identity $<\alpha, \beta><\beta, \alpha> = 4\cos^2\theta$.

Root systems	System Graph	Coxeter Graph
$<\alpha, \beta><\beta, \alpha> = 0, \theta = \pi/2$ $A_1 \times A_2$	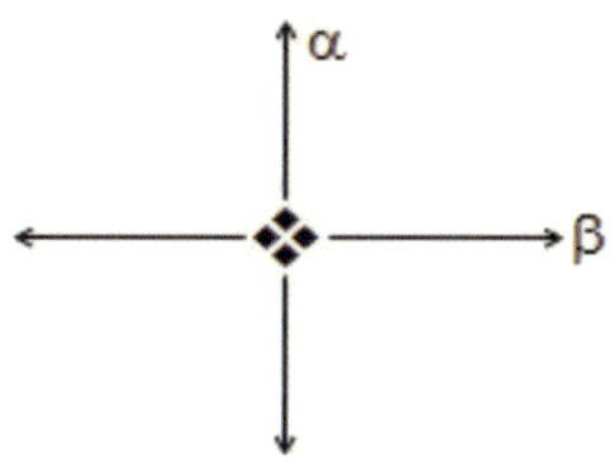	
$<\alpha, \beta><\beta, \alpha> = 1, \theta = \pi/3$ A_2	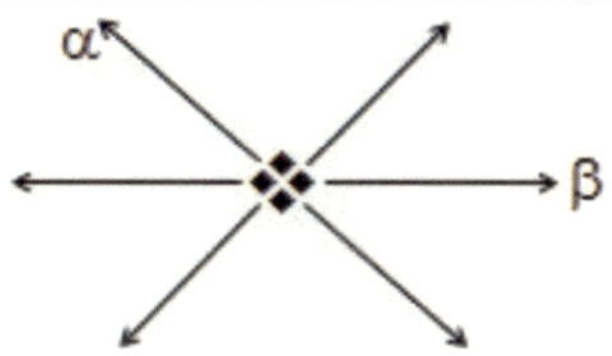	
$<\alpha, \beta><\beta, \alpha> = 1 \bullet 2 = 2, \theta = \pi/4$ B_2	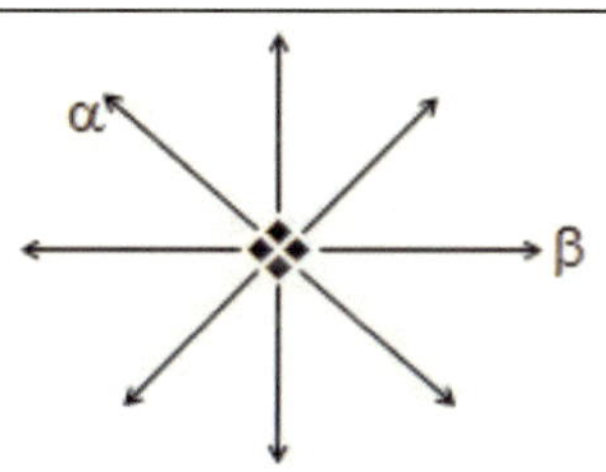	
$<\alpha, \beta><\beta, \alpha> = 1 \bullet 3 = 3, \theta = \pi/6$ G_2	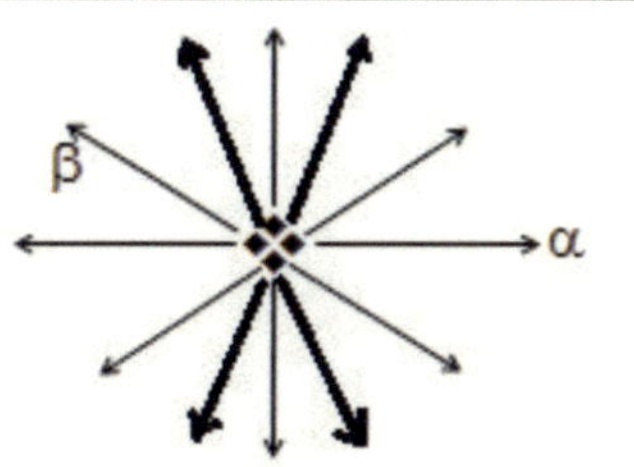	

Table 1. Strings and their root spaces

Def. D. 3. A Coxeter graph of Φ, is the graph of l, vertices which the ith-vertex is joined to the jth-vertex with $i \neq j$, for $\langle \alpha_i, \alpha_j \rangle = \langle \alpha_j, \alpha_i \rangle$, edges. For example, to the root systems of range $l = 2$:

The Coxeter graph determine the numbers $\langle \alpha_i, \alpha_j \rangle$, in the case of that all roots have the same length. Then $\langle \alpha_i, \alpha_j \rangle = \langle \alpha_j, \alpha_i \rangle$. In a more general case, where there is more of one different length (case of the root systems of range two; G_2, B_2), the graph fails to the concrete case in which is had a vertices pair that correspond to a simple short root, which cans be a long root or of big length.

A relevant fact on the Coxeter graph is that these determine completely to Weyl group, essentially because these determine the orders of the direct products of generators of W.

Proposition D. 2. From the Table 1, of inner products and length of roots that change in the interval $0 \leq \langle \beta, \alpha \rangle \langle \alpha, \beta \rangle \leq 4\cos^2\theta$, is had that the order of $\sigma_\alpha \sigma_\beta$, in W, is respectively 2, 3, 4, 6, when $\theta = \pi/2$, $\pi/3$ (or $2\pi/3$), $\pi/4$ (or $3\pi/4$), $\pi/6$ (or $5\pi/6$). [Note that $\sigma_\alpha \sigma_\beta$, is the rotation through 2θ].

To the case when we have roots of short length, is ordinary to add an arrow to designate it in the Coxeter graph. This additional information help us to recover the integers of Cartan, called to resulting graph; Dynkin diagram of Φ (as before, this depend on the number of simple roots). Example;

B_2 : 

G_2 : 

Somehow someone exist such component irreducibles like the system of roots shaped by a pair of roots. Let's study the component irreducible systems. Let's remember that a system Φ, is irreducible if and only if, Φ(or equivalently Δ) do not admit a division into two orthogonal proper subsets. Then by the correspondence between Coxeter graphs and root systems, is clear that Φ, is irreducible if and only if their graph of Coxeter is connect (in the usual sense)[25].

In general is possible to give a number of connect components of the Coxeter graph, likewise, let

$$\Delta = \Delta_1 \cup \Delta_2 \cup \Delta_3 \cup \ldots \cup \Delta_l, \tag{D. 7}$$

the corresponding partition of Δ, in mutually orthogonal subsets. If E_i, is the generated by Δ_i, is clear that

[25] If this is non-connect it would be equivalent to that f, admits a partition into two orthogonal proper subsets.

$$E = E_1 \oplus E_2 \oplus \ldots \oplus E_l, \tag{D. 8}$$

For other side, the linear $\mathbb{Z}$-combinations of Δ_i, which are roots (the root system Φ_i) obviously shape a root system in E_i, whose Weyl group is the restriction to E_i, of all σ_α, $\forall$ $\alpha \in \Delta$, that is to say, the W-orbit of the root $\alpha \in \Delta$, if and only if σ_α, acts trivially in E_i.

Proposition D. 3. Let $E' \subset E$. If the reflection σ_α, leave E' invariant then $\forall$ $\alpha \in E'$, $E' \subset P_\alpha$, with $P_\alpha = \{\beta \in E \mid (\beta, \alpha) = 0\}$.

Note: Demonstrating this proposition is deduced immediately that every root fall in some of the spaces E_i, where

$$\Phi = \Phi_1 \cup \Phi_2 \cup \Phi_3 \cup \ldots \cup \Phi_l, \tag{D. 9}$$

with $1 \leq i \leq l$.

Proof. Let $E' \subset E$. Then to any $\beta \in E'$, then $\beta \in E$, and $\sigma_\alpha(\beta) = \beta \in E'$, that is to say, $\sigma_\alpha(E') = E'$, but as $E' \subset E$, then $\sigma_\alpha(E) = E$, and $\sigma_\alpha(E') \subset \sigma_\alpha(E)$. But

$$\sigma_\alpha(E) = \{\beta \in E \mid (\beta, \alpha) = 0\} = P_\alpha,$$

and since $\sigma_\alpha(E') = E'$, then

$$\sigma_\alpha(E') = E' \subset \sigma_\alpha(E) = P_\alpha,$$

where $E' \subset P_\alpha$. ■

Then to every $\alpha_i \in E_i$, $1 \leq i \leq l$, and $\forall$ $\alpha_i \in \Phi$, is had that (D. 9).

Proposition D. 4. Φ, is decomposed (univocally) as the union of irreducible root systems Φ_I (in subspaces E_i, of E) such that

$$E = E_1 \oplus E_2 \oplus \ldots \oplus E_l$$

Proof. One simple consequence of the before proposition, since if every $E_i \subset P_\alpha$ (that is to say, $\alpha_i \in E_i$) with $1 \leq i \leq l$, then to

$$\sum_{\gamma_i \in \Delta_i} < \alpha_i, \beta > \gamma_i = 0 (\Delta_i \subset \Delta), \tag{D. 10}$$

is had that $<\alpha_i, \beta> = 0$, thus $E = E_1 \oplus \ldots \oplus E_l$. ■

The before discussion establish a criteria of enougness to classify irreducible root systems or equively the Dynkin diagrams, this last, by the proposition that extends bijections between Euclidean root spaces to an isomorphism between corresponding root systems to said

Euclidean spaces, which are isomorphic in nature form. The isomorphism will be linearly represented by the Cartan matrix in every case.

Theorem D. 1. If Φ, is a irreducible root system of range l, their Dynkin diagram is one of the following cases to l-vertices:

Group	Dynkin Diagram
A_l ($l \geq 1$)	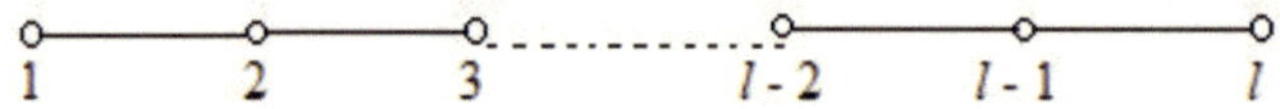
B_l ($l \geq 2$)	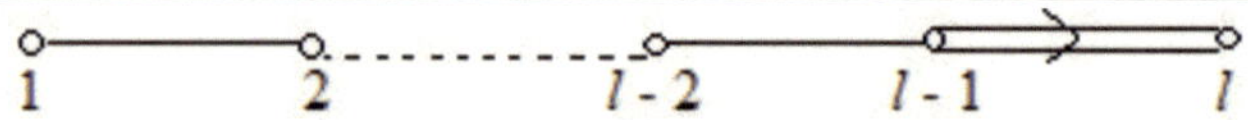
C_l ($l \geq 3$)	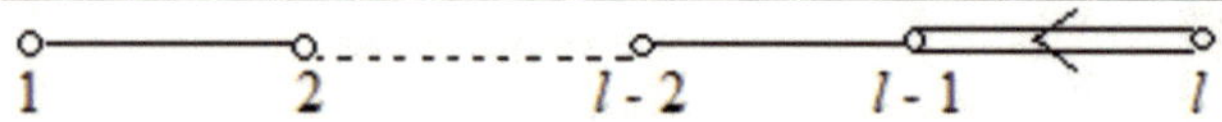
D_l ($l \geq 4$)	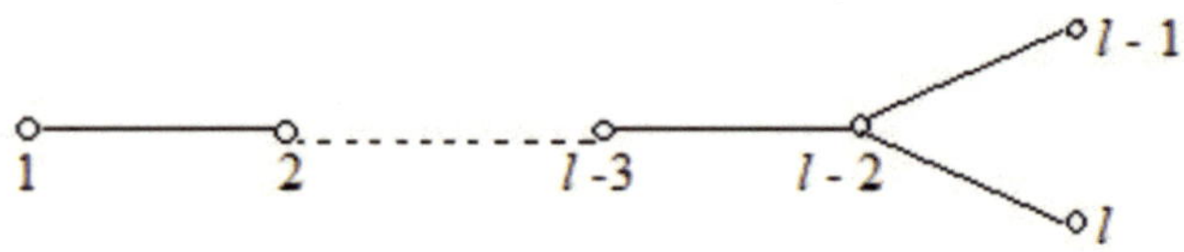
E_6	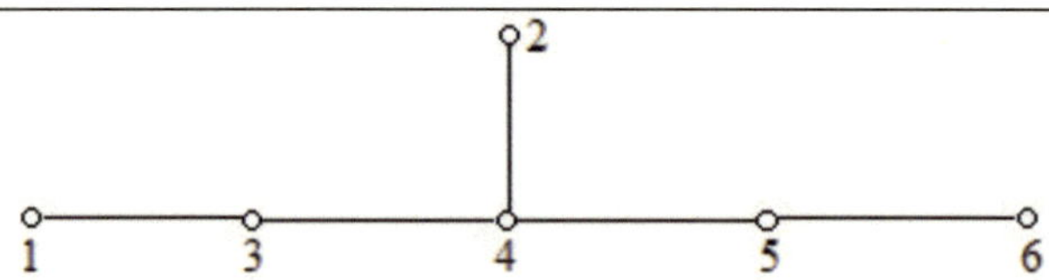
E_7	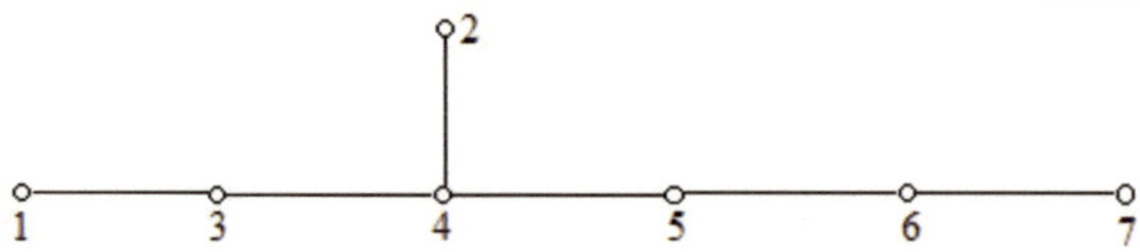
E_8	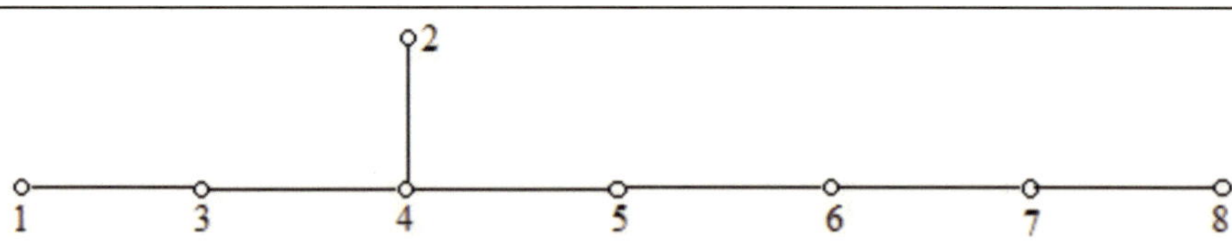
F_4	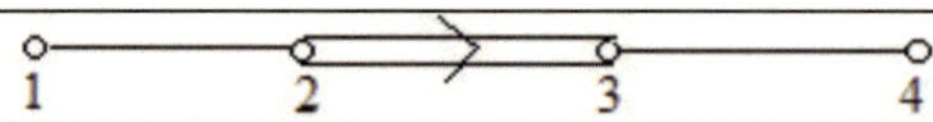
G_2	

Note: The signed restrictions to the integer l, that is to say, $l \geq n$, $\forall\ n \in \mathbb{Z}$, are considered only to not duplier before cases.

Table 2. Coxeter Groups and their Dynkin graphs.

APPENDIX E: Definition of a Real Reductive Group

Let G, be a semisimple Lie group. A real reductive group is a complex algebraic group defined on the real numbers, which is covering of an open subgroup of a group of real points (such real points are Cartan subalgebras of the Lie real reductive algebra).

Let $G_{\mathbb{C}}$, is a symmetric subgroup of GL(n, $\mathbb{C}$), with real points and let the real connected component $G_{\mathbb{R}}$, of $G_{\mathbb{C}} \cap GL(n, \mathbb{C})$; a real reductive group $\mathbb{G}$, is the finite covering of an open subgroup G_0, of $G_{\mathbb{R}}$. Of this way, if p is the covering homomorphism defined for the map

$$p: G \rightarrow G_0 \subset G_{\mathbb{R}}, \tag{E. 1}$$

whose rule of correspondence is

$$g \mapsto p(g), \tag{E. 2}$$

then explicitely a real reductive group is the space

$$\mathbb{G} = \{\mathbf{g} = p(g)^{-1} \in G \mid p(g) \in G_0 \subset G_{\mathbb{R}} = G_{\mathbb{C}} \cap GL(n, \mathbb{C})\}. \tag{E. 3}$$

Exercises

1. Consider the cohomology of R_{∂}. Demonstrate that R_{∂} on a Koszul complex of dimension *n* is a *n*-dimensional John integral.
2. Demonstrate that the K-orbits in a flag manifold X = G/B, where B is the vector bundle of Borel subalgebras of $\mathfrak{g}$ = Lie(G), are naturally parameterized by the right classes W_K/W_G of the Weyl group of G.
3. Demonstrate that if s_μ, is defined by $s_\mu H = H - B(h, H)H$, to $H \in \mathfrak{a}$, then $Ad(k)H = s_\mu H$.
4. Be $\mathbb{P}(S^+) \cong \mathbb{CP}_7$. We define the orbit

$$\mathbb{P}^+ = \{[z] \in \mathbb{CP}_7 | \in(z, z) = ir, \forall\, r > 0\}.$$

This is a complex orbital submanifold of $\mathbb{P}$. A compact maximal submanifold in $\mathbb{P}$, is a copy of $\mathbb{CP}_3$, having complex dimension three. Demonstrate that the Penrose transform $\mathcal{P}$, on the cohomological classes $H^3(\mathbb{P}^+, \mathcal{L})$, give the isomorphism of right fields:

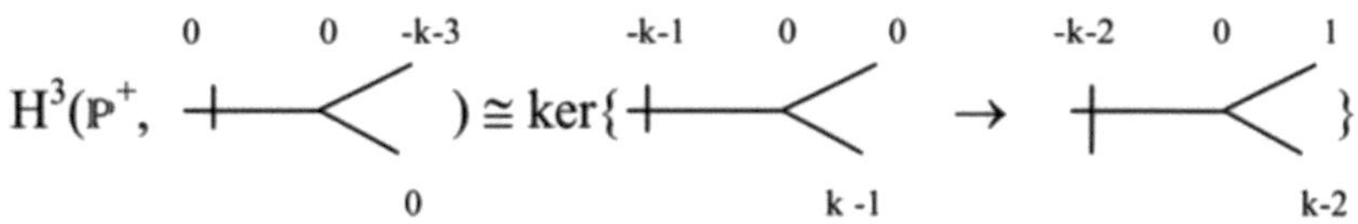

where $\mathcal{L}$, is a homogeneous bundle of lines in $\mathbb{P}$, where $k \geq 3$.

5. Demonstrate that traditional contours (or classic) given by $H_{f+d}(\Pi - \Upsilon, \mathbb{R})$, we can be visualized like cohomological functionals of co-cycles of Λ, in $H^{f+d}(\Lambda - \Sigma)$, with Σ, a flag manifold.
6. Consider the cohomological group $H_8(\Pi - \Upsilon', \mathbb{C}) = \mathbb{C}$, and demonstrate that the image of the generator of this under group of the map of Mayer-Vietoris is a physical contour usual to the inner product.
7. Consider the theory of the mother gravity (Ramírez, F. M., Ramírez, F. L., Appliedmath II/IPN, 2006) given by the glues of cells in a topological space X. Determine a state $\phi \in H^2(\mathbb{P}_1 \times \mathbb{P}_2 \times \mathbb{P}_3 \times \mathbb{P}_4, O(-2, -2, -2. -2))$, that determine the domain of the coordinate of gravity on the others coordinates of state. How we give a cohomological re-interpretation of the integral formula of type Cauchy $\forall\ Z^\alpha Z^\beta f_{-6}(Z)\Omega$, to the "gravitational charges" in the mother gravity?

Use the diagram: 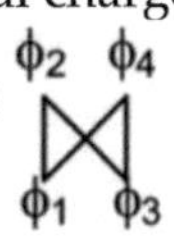

INTECH
open science | open minds

8. Demonstrate that the aseveration of the exercise 6, is equivalent to the asseveration of that exist a only cohomological contour to the inner product and suggest a method to demonstrate that those contours are in effect, cohomological.
9. Determine the Penrose transform on:

a. The quadric $\mathbb{M}$, of dimension six and projective space $\mathbb{P}$, of the 3- dimensional α–planes in $\mathbb{M}$, with their corresponding Bernstein-Gelfand-Gelfand resolution to the canonical bundles on $\mathbb{P}$.

b. On the flat semi-conformal complex spinorial 4-manifold **M**.

c. On a complex line of the projective space $\mathbb{P}^3(\mathbb{C})$, and of the bundle of lines $O(-2)$.

d. On the co-cycles of the $\mathfrak{u}$-specialization of the space $H^{0,q}(G/L, V)$.

e. Whose integral cohomology of vector fields is $H^1(\mathbb{B}^{-+-}, O(-4, 3))$.

Here $\mathbb{B}^{-+-}$, is the open orbit or minimal K-to the action of the group SU(1, 2), on $\mathbb{B}$. This conform the space

$$\mathbb{B}^{-+-} = \{L_1 \subset L_2 \mid \Phi|_{L1}, \text{ is defined positive and } \Phi|_{L2}, \text{ have type } (+\,-)\}.$$

wth Φ, a Hermitian form on $\mathbb{C}^3$, of type $(+\,-\,-)$.

10. Which is the Penrose transform of the holomorphic succession of De Rham on $\mathbb{P}$, given by $0 \to \mathbb{C} \to \Omega^{\bullet}$, that include all the spaces of fields that are of interest in electromagnetic theory: potentials module gauge and fields for both parts of positive and negative frequency?
11. Demonstrate that the integrals of twistor electrical field

$$H^1_{\mathcal{L}}(U'', O/\mathbb{C}) \to H^2_{\mathcal{L}}(U'', \mathbb{C}) \cong \mathbb{C},$$

and

$$H^1_{\mathcal{L}}(U'', O(-4)) \to H^4_{\mathcal{L}}(U'', \mathbb{C}) \cong \mathbb{C},$$

cans be re-write using resolutions of $\odot^{n-2}T^*$, and $\odot^{n-2}T$, like

$$H^1_{\mathcal{L}}(U'', O(n-2)) \to H^4_{\mathcal{L}}(U'', \odot^{n-2}T^*) \cong \odot^{n-2}T^*,$$

and

$$H^1_{\mathcal{L}}(U'', O(n-2)) \to H^4_{\mathcal{L}}(U'', \odot^{n-2}T) \cong \odot^{n-2}T.$$

12. Demonstrate through of classical functionals that in a contour cohomology, the singularities of a complex space are poles of Cauchy.
13. Demonstrate using Radon transform of dimensions, that an osculating space of order p, of a submanifold of dimension m, have dimension

$$m + C(m + 1, 2) + \ldots + C(m + p - 1, p).$$

14. Demonstrate that the integral $\int \phi(z)f(z)dz$, with $\phi(z) \in \mathbb{C} \setminus F$, and $f(z) \in C^{\infty}(F)$, is a hypercomplex generalization of the integral of Cauchy.
15. Using arguments of *MHD*, in the universe and considering that the orientation of the spiral arms of the galaxies are oriented by intersidereal magnetic fields, demonstrate that all the universe is a conductor fluid.

[Suggestion: Use topological arguments of connectivity and integration in chains to satisfy the Helmholtz theorem and other necessary results in the process of materialization of the universe like a fluid. Demonstrate that the universe have many degrees of circulation representing by cohomologies of vector fields isomorphic to contour cohomologies]

16. Let M, be a complex Riemannian manifold of order of range p, and compact then the integral $i/(2\pi)\int_M \omega_{p,\,p+1} \wedge \underline{\omega}_{p,\,p+1}$, is a integer.
17. In L^2-cohomology: Be $H_{\sigma,\,\mu}$, the $(\mathfrak{p}, {}^0M)$-module with $\mathfrak{a}$, acting by $(\mu + \rho)I$, and $\mathfrak{n}$, acting by 0. Be V, a $(\mathfrak{g}, K)$-module. If $T \in Hom_{\mathfrak{g},K}(V, H^{\sigma,\mu})$, then $T^{\sim}(v) = (v)(1)$, $\forall\, v \in V$, is to say, $T^{\sim} \in Hom_{\mathfrak{p},\,0M}(V/\mathfrak{n}V, H_{\sigma,\,\mu})$. You enunciate a orbital integral belonging to a class of $V/\mathfrak{n}V$. Is the operator $T^{\sim}$, a intertwinning operator?
18. Enunciate three extensions of $(\mathfrak{g}, K)$-modules in the class of the L^2-modules. Which is the image of the functor *Ext* of the group of cohomology $H^{\bullet}(,(\mathfrak{m}/\mathfrak{t}), V \otimes \mathfrak{t})$?
19. Determine a hyperbolic minitwistor space of $SL(4, \mathbb{C})$. Give their physical interpretation to solutions of the electromagnetic wave equation.
20. If M, is a inferiorly bounded orientable compact being also a pseudo-Riemannian manifold with the characteristic metric g, with indices Ind(g), then d and δ are adjuncts.

Suggestion: Consider the differential forms $\alpha \in \Omega^k(M)$ and $\beta \in \Omega^{k+1}(M)$ and demontrate that

$$<d\alpha,\beta> = \int d\alpha \wedge_* \beta = \int \alpha \wedge_* \delta\beta = <\alpha,\delta\beta>$$

21. Demonstrate that in the solution space of the differential equations corresponding to the cohomology $H^1(|\mathbb{P}^{*-}|, \mathfrak{O})$ (dual twistor space of $H^1(|\mathbb{P}^{+}|, \Omega^3)$) the twistor transform preserve L^2.
22. Compute all the unitary representations of SU(2, 2).
23. Give a example of unitary representations used in the representation of the tensor of curvature $R_{\alpha\beta}$, in function of spinor matrices.
24. Why is important the concept of real or complex analytic null curve in twistor geometry to the description of strings or minimal surfaces in M, or M?
25. Which are the real and complex hyperbolic orbits on the which we can compute integrals that conform a L^2-cohomology?
26. You obtain the solutions through of the hyperfunctions of the wave equation

$$\Box^2 \phi = 0,$$

in the space $\mathbb{R}^{N+1}$.

27. Using complex lines in the space $\mathbb{P}^1(\mathbb{C})$, determine the isomorphic space to $\mathbb{P}^1(\mathbb{C})$ whose realization in $\mathbb{P}^5(\mathbb{R})$, is a null quadric. Whose null quadric represents a twistor hypersurface of the action of the Lie group $O_o(5, 1)$, on $\mathbb{P}^3(\mathbb{C})$.
28. Demonstrate that a L^2-curvature is a generalized curvature in a symmetrical homogeneous space with positive defined Hermitian form.
29. Compute the Penrose transform on:
i. The cycles of the flag manifold of complex dimensional (2, 3).
ii. On holomorphic pencils of a surface of 2-dimensional Lobachevski.
iii. On a α-curve of a analytic hypersurface of a complex Riemannian manifold of dimension 4.
iv. On the sheaf $O(\odot T^2\mathbf{M})$, of germs that are quadric forms in **M**.
v. Strings of a 10-dimensional p-brane.
30. Generalize the integral of the monopole

$$g(z) = (1/2\pi i)\int f(z)/(z - z_0)\, dz,$$

to the case $\mathbb{C}^2$, considering circles S^1, like orbits in $\mathbb{C}^2$.

31. Give an example of a integral curvature on symmetrical spaces that conforms a L^2-cohomology on a real Riemannian manifold and whose tensor of curvature be the direct sum of regular images of co-cycles of the real Riemannian manifold in question.
32. Let $\Omega_H = dv^b + \eta\nabla^2\Omega_H \in \mathcal{I}I_H(\mathbf{E}, \sigma)$, be with ^{b}, Euclidean in M^3, (is to say in $\mathbb{R}^3$) and with materials of $\mathcal{I}I_H$, that satisfies $\partial\rho/\partial t + \mathrm{div}(\rho\mathbf{v}) = 0$. Demonstrate that the equation of the vorticy currents take the form of the equation of the "frozen fields",

$$d/dt(\mathbf{H}/\rho) = (\mathbf{H}/\rho \bullet \mathrm{grad})\, \mathbf{v},$$

Use the orbits of the classes space given by $G/C(\mathbb{T})$, with $\mathbb{T}$, a complex torus.

33. Calculate the integral on cycles of the hyperbolic 2-dimensional disc D^2, that are generalized orbits of D^2.
34. Let G, be a Lie group; H, and N, closed subgroups such that

$$H \subset N \subset G.$$

Assume that G/H, and G/N, have G-invariant positive measures dg_H, and dg_N, respectively. Demonstrate that N/H, have a N-invariant positive measure dn_H, the which (normalized adequately) satisfies

$$\int_{G/H} f(gH)\, dg_H = \int_{G/N}\left(\int_{N/H} f(gnH)\, dn_H\right) dg_N,$$

$\forall\, f \in C_c(G/H)$.

35. Which are the orbital integrals to the formal integral of Feynman type given in a evaluating given by

$$(\Gamma,\Omega) = \sum_i \int_{C_\Gamma^0(\psi)} \omega(\Gamma))_i = \int_{\Gamma=\Gamma_1\cup\Gamma_2} \varphi = \int_{\Gamma_1} \varphi(e) + \int_{\Gamma_2} \varphi(e) = 0,$$

where Γ_j, are the jth-paths in the evaluation of the integral on Jacobi graphs space $C_\Gamma^0(\psi)$. ψ, is the model of the graph used to describe the path or trajectory used by the electron e, with energy state $\phi(e)$.

36. Using complex vector bundles of lines classify the differential operators used in the elliptic differential equations until second order.
37. Give a Intersection cohomology to the integral operators evaluated in singular spaces. For example, knot, whole, cuspidal space, etc.
38. Demonstrate that the velocity field of the bubble produces the deformation of the bubble in movement within $\tilde{\ } | |$. In fact it will be the cause of all the phenomena associated in the bubble in movement. Use a topological scheme.
39. Demonstrate that the Cauchy tensions as well as the tension stresses over the bubble elastic surface are produced by the fluid viscosity tensions. Use a topological scheme.
40. Demonstrate the identity of dimension theory

$$\dim \mathfrak{M}_t = \int_G \dim B^{\dim\chi(g)} \mathrm{codim}\, B^0\, \mu\Lambda(g),$$

where $\Lambda(g)$, is a measure of Borel positive define

$$\dim \mathbb{G} = \dim |B| \aleph_0^{\aleph},$$

where $\aleph_0$, is the first infinite cardinal in set theory. $B^0 = B/\mathbf{1}\cdot$, where

$$\mathbb{B} = \{B_i \mid 0 \leq \mu(B_i) \leq 1\},$$

41. Consider $V \in \mathcal{V}$. Then

1. $V = \oplus V_\mu$, $\dim V_\mu < \infty$, $\forall\ \mu \in (\mathfrak{a}_C)^*$,

2. there are $\Lambda_1, \ldots, \Lambda_q \in (\mathfrak{a}_C)^*$, such that $V_{\Lambda j}$, is not vanishing and if V_μ, is not vanishing then $\mu = \Lambda_j - Q_j$, to some $j \leq q$, and $Q \in \mathbb{L}^+$.

Note: $V_\mu = \{v \in V \mid (H - \mu(H))^k v = 0.$ t. s. $k \in \mathbb{Z}^+$, and $H \in \mathfrak{a}\}$, and the space $\mathbb{L}^+$, is the set of points:

$\mathbb{L}^+ = \{$Set of the all integer combinations of non-negative elements of $\Phi(\mathfrak{a}, \mathfrak{g})\}$,

Also the space $\mathcal{V}$, is the category

$$\mathcal{V} = \{\text{Category of finitely generated } (\mathfrak{g}, {}^0M)\text{-modules } V\}.$$

42. Consider the representation (π, H), of integrable square such that $\forall\ g \in G$, $w \in H_K$, and $v \in H^\infty$, is satisfied

$$\int_G |\langle \pi(g)v, w\rangle|^2 dg < \infty,$$

Demonstrate that $\forall\ a \in A$,

$$\int_A (1 + \log\|a\|)^{d-r} da < \infty,$$

with $\pi(g) \in L^2(G)$, if and only if $r \to \infty$.

43. Construct an intertwining integral operator to the representations $Ind_{MAN}{}^G(1 \otimes e^{\rho L} \otimes 1)$, and ind $A(\underline{P}, P, 1, \rho_G)$, where $\rho_L = \alpha/2$, (α, is a restrict root in L) with λ, such that $\lambda + 2\delta(\mathfrak{u} \cap \mathfrak{p}) = 0$. Also

$$\underline{P} \subset P,$$

with $P = MAN$, and $\underline{P} = MA\underline{N}$, with $N = \theta N$, where θ, is an involution. $\rho_G = \Sigma_G{}^+$, is the set of restrict roots of G.

44. Demonstrate that the operator of Szego type

$$S : ind_{MAN}{}^G(\sigma \otimes e^{\rho L} \otimes 1) \to ind_L{}^G(V^{\pi'}) \hookrightarrow ind_L{}^G(C^{\#}{}_{\lambda} \otimes (\wedge^s \mathfrak{u})*),$$

should be given by on integral formula of the form

$$Sf(x) = \int_{L/L \cap MAN} \pi'(l)[Tf(xl)]dl = \int_{L \cap K} \pi'(l)[Tf(xl)]dl,$$

With $T : V^0 \to V^{\pi'} \subset C^{\#}{}_{\lambda} \otimes (\wedge^s \mathfrak{u})*$, an $L \cap M$, map.

45. From the before problem (problem 44), What happens when M is compact?
46. Use the definition of the Harish-Chandra function Ξ, to demonstrate that the finite integral

$$I(t) = \int_{\underline{N}F} a(n)^{\rho F}\ \Xi_F(\underline{n})(1 - \rho(\log(\underline{n}))^{r-q} d\underline{n},$$

$\forall\ q > d + r$, take the form

$$I(t) = \int_{\underline{N}F} a(n)^{\rho F} \int_{\underline{N}F} u(km_F(\underline{n}))^{\rho} dk (1 - \rho(\log a(\underline{n}))^{r-q} d\underline{n}.$$

47. Use the Harish-Chandra function to demonstrate that if (π, H), is of square integral then (π, H), satisfies the strong inequality.
48. Let σ_K, be a representation of $P/{}^0M = AN$, in $V/\mathfrak{n}^k V/{}^0\mathfrak{m}$. Let $I_K : V/\mathfrak{n}^k V \to V/\mathfrak{n}^k V$, be and we consider $I_K(V/\mathfrak{n}^k V) = I^{\wedge}{}_K = T_K$. Demonstrate that $\ker T_K$, contains to $\mathfrak{n}^k V$.
49. Which are the *D*-modules that agree with the representations given by images under Penrose transform?
50. How could be related the Radon transform on $D_{G/H}$-equivariant modules with the Penrose-Ward transform?
51. Let G, be a (separable, metrizable) locally compact group and H, a closed subgroup of G. Suppose that G, and H, are unimodular, and let m_G, and m_H, denote Haar measures on G, and H, respectively.

Let μ, be a measure of positive type on H. If μ, is identified with their image under the canonical injection of H, in G. Show that μ, is of positive type as a measure on G. (For each function $f \in \mathcal{H}(G)$, show that

$$\iint_{G\times H} f(sx)\,\overline{f(s)}\,dm_G(s)d\mu(x) = \iiint_{G\times H\times H} h(s)f(sy^{-1}x)\,\overline{f(sy^{-1})}\,dm_G(s)dm_H(y)d\mu(x),$$

where $h \in \mathcal{H}(G)$, is such that

$$\int_H h(su)dm_H(u) = 1,$$

52. Let $H_{n,\Gamma}$, denote the space of automorphic forms f, on P, which are holomorphic on P, and such that for each $\gamma \in \Gamma$, we have

$$f(z) = J_\gamma(z)^n f(\gamma \bullet z),$$

where $J_\gamma(z) = (cz + d)^{-1}$, if

$$\gamma = \begin{pmatrix} a & b \\ c & d \end{pmatrix},$$

Since $\gamma \bullet z = (-\gamma)\bullet z$, we cannot have f = 0, unless $n = 2k$, is even; in which case f, is said to be an automorphic form of weight k, relative to Γ. Let f, be any holomorphic function on P, and $F = f^{z_0}$. Show that for each compact subset A, of G, there exists a compact neighborhood B, of A, in G, such that

$$\sup_{s\in A}|F(s)| \le M\int_B |F(s)|\,dm_G(s),$$

Where M, is a constant independent of f. (Apply Cauchy's formula to f, and the projection of A, on G/K = P). Deduce that if $F \in H^1(\alpha_n)$, the family $(\sup_{s\in A}|f(\gamma s)|)_{\gamma\in\Gamma}$, is summable (use the last that $BB^{-1} \cap \Gamma$, is finite). Consequently the family $\{f(\gamma s)\}_{\gamma\in\Gamma}$, is absolutely summable for each $s \in G$, and the function

$$F_\Gamma(s) = \sum_{\gamma\in\Gamma} F(\gamma s),$$

is of the form f^{z_0}, where f_Γ, is an automorphic form of weight k, on P, which is holomorphic in P.

53. Let G, be a (separable, metrizable) locally compact commutative group. For each continuous unitary representation U, of G, on a separable Hilbert space E, show that there exists a unique representation L, of the involutory Banach algebra $\mathscr{B}_{\mathbb{C}}(G^\wedge)$ (with thse usual product) on E, such that $L(\langle s, \bullet\rangle) = U(s)$, for all $s \in G$.

54. Let (G, K), be a Gelfand pair and let g, be a function belonging to $\mathscr{C}(K\backslash G)$, such that for each $f \in \mathscr{H}(K\backslash G/K)$, there exists a complex number λ_f, such that $f * g = \lambda_f g$. If $g(s_0) \neq 0$, for some $s_0 \in G$, show that the function

$$\omega(s) = g(s_0)^{-1} \int g(sts_0)dm_K(t),$$

is a spherical function.

55. Let G, be an amenable group, let K, be a compact subset of G, and M, a real number major that 0. Show that for each $\varepsilon > 0$, there exists a function $f \geq 0$, belonging to $L_{\mathbb{C}}^1(G)$, such that $N_1(f) = 1$, and such that, for each function $g \in L_{\mathbb{C}}^1(G)$, which is zero on K, and satisfies $N_1(g) \leq M$, we have

$$N_1((g * f) - (\int g(x)dx\alpha|_f) \leq \varepsilon,$$

where α, is Haar measure on G.

56. Demonstrate that $H^{1,0}{}_{[\mathbb{P}^1/\infty]}(\mathbb{P}^1, \mathcal{O}(n))$, is not a Verma module to $n \geq 2$.

57. Let (G, K), be a Riemmanian symmetric pair of the non-compact type and assume that G, has finite center. Let

$$E_f(a) = e^{\rho(\log a)} \int_N f(an)dn,$$

$\forall f \in L^{\#}(G)$, such that

$$\int_G \phi_v(x)f(x)dx = \int_G e^{iv(\log a)} F_f(a)da,$$

Verify the following assertions:

i. $F_{f*g} = F_f * F_g$, $\forall$ $f, g \in L^{\#}(G)$,
ii. $F_{f^*} = (F_f)^*$, $\forall$ $f \in L^{\#}(G)$,
iii. The mapping $f \mapsto F_f$ $(f \in L^{\#}(G))$ is one-to-one.

Note: $L^{\#}(G) = \{f \in L^1(G) \mid f \text{ are bi-invariant under } K\}$.

58. Let G, be a Lie group, H and N closed subgroup such that $H \subset N \subset G$. Assume that G/H, and G/N, have positive G-invariant measures dg_H, and dg_N. Show that N/H, has an N-invariant positive measure dn_H, which (Suitable normalized) satisfies

$$\int_{G/H} f(gH)dg_H = \int_{G/N} \left(\int_{N/H} f(gnH)dn_H \right) dg_N,$$

$\forall f \in C_c(G/H)$.

59. Let G, be a semisimple connected compact Lie group and let T, be a maximal torus in G. Let w denote the order of the Weyl group of G, and let

$$D(t) = \prod_{\alpha\in\Delta^+} 2sen\left(\frac{1}{2}\alpha(iH)\right),$$

if t = expH∈T. Let dt, and dg, respectively denote the invariant measures on T, and G, normalized by

$$\int_T dt = \int_G dg = 1,$$

Derive Weyl's formula

$$\int_G f(g)dg = \frac{1}{w}\int_T |D(t)|\,dt \int_G f(gtg^{-1})dg,$$

∀ f∈C(G).

60. Let V, be a n-dimensional vector space over a non-archimedean local field F, for instant the field of p-adic numbers. Let γ: V → V, be a linear endomorphism with distinct eigenvalues in an algebraic closure of F. The centralizer I_γ, of γ, is of the form

$$I_\gamma = E_1^\times \times \ldots \times E_n^\times,$$

where $E_1^\times$, ..., $E_r^\times$, are finite separable extensions of F. This is a commutative locally compact topological group. Let $\mathcal{O}_F$, denote the ring of integers in F. We consider the set of lattices of V, that are sub-$\mathcal{O}_F$-modules $\mathcal{V}\subset$ V, of finite type with maximal rank. We are interested in the subset $\mathcal{M}_\gamma$, of lattices $\mathcal{V}$, of V such that $\gamma(\mathcal{V}) \subset$ V. The group I_γ, acts the set $\mathcal{M}_\gamma$. This set is infinite in general but the set of orbits under the action of I_γ, is finite. We fix a Haar measure dt, on the locally compact group I_γ. We consider a set of representatives of orbits of I_γ, on $\mathcal{M}_\gamma$, and for each x in this set, let denote $I_{\gamma,x}$, the compact open subgroup of I_γ, of elements stabilizing x. Demostrate that the finite sum

$$I = \sum_{x\in\mathcal{M}_\gamma/I_\gamma} \frac{1}{Vol(I_{\gamma,x}, dt)},$$

is a typical orbital integral.

61. How is constructed the orbit of the holomorphic vector bundle seated in homogeneous space G/L, with L, compact locally?
62. Suppose that $f * \omega = \lambda_f\omega$, ∀ f∈$\mathcal{K}$(K\G/K), since ω(e) = 1, with G, unimodular. Then

$$\lambda_f = (\tilde{f} * \omega)(e) = \int_G f(s)\omega(s)dm_G(s) = \langle f, \omega\rangle.$$

63. The product of two functions of positive type on G, is a continuous function of positive type.
64. Let G, an unimodular locally compact group (separable and metrizable) and K, a compact subgroup of G. Let m_K, be the Haar measure on K, with total mass equal to 1. If we put

$$f^{\#}(s) = \iint_{K\times K} f(tst')dm_K(t)dm_K(t'),$$

For all function $f \in \mathcal{K}(G)$, the mapping $f \mapsto f^{\#}$, is a projector on the vector space $\mathcal{C}(G)$, onto the vector space $\mathcal{C}(K\backslash G/K)$. Let prove $\forall\ f \in \mathcal{C}(K\backslash G/K)$, and $g \in \mathcal{C}(G)$, that

$$(fg)^{\#} = fg^{\#}.$$

65. Let $K \subset \mathbb{R}^n$, be a symmetric compact convex set having 0, as an interior point. Let V, denote the Lebesgue measure of 2K = K + K. Show that if 0, is the only point of $\mathbb{Z}^n$, which belongs to 2K, then

$$2^n = V + \frac{1}{2^{2n}V} \sum_{\substack{m \in \mathbb{Z}^n \\ m \neq 0}} \left| \int_K e^{-2\pi i(x)mi} dx \right|^2.$$

(Show that Poisson's formula (with $G = \mathbb{R}^n$, and $H = \mathbb{Z}^n$) may be applied to the function $f = \varphi_K * \varphi_K$). Hence give another proof of Minkowski theorem.

66. Considering the Cartan matrix

$$\begin{bmatrix} 2 & -1 & 0 & 0 \\ -1 & 2 & -2 & 0 \\ 0 & -1 & 2 & -1 \\ 0 & 0 & -1 & 2 \end{bmatrix},$$

of the root system $\Phi = F_4$, let obtain the graph if $\|\beta\| < \|\alpha\|$.

67. Let define the equivalences given by the Penrose-Ward transform of the twistor correspondence

$$\begin{array}{ccc} & \mathbb{C}^4 \times \mathbb{P}^1 & \\ \swarrow & & \searrow \\ \mathcal{P} & & \mathbb{C}^4, \end{array}$$

68. What about topological B-model and twistor strings?
69. What is a quadric in $\mathcal{P}^{2|3} \times \mathcal{P}^{2|3}$? Which are their orbits?

70. Let a >0, b > 0, and let f, be the continuous real-valued function defined on $\mathbb{R}$, which is equal to b, when x = 0, is zero for |x| ≥ a, and is linear in each of the intervals [–a, 0], and [0, a]. We have

$$(F \;\; f)(t) = \frac{b \sin^2 \pi at}{\pi^2 at^2},$$

Deduce that if

$$g(x) = \sum_{n=-N}^{N} f(x+n),$$

Then

$$(F \;\; g)(t) = \frac{b \sin^2 \pi at}{\pi^2 at^2} \frac{\sin(2N+1)\pi t}{\sin \pi t}.$$

71. Let give the Dynkin diagram for a complex simple Lie algebras:

a. $\mathfrak{so}(7, \mathbb{C})$,

b. $\mathfrak{sl}(4, \mathbb{C})$,

72. Which are the Dynkin diagrams to flag manifolds of lines planes in $\mathbb{C}^4$, given by $\mathbb{F}_{12}$, and $\mathbb{F}_{123}$, from the corresponding matrix $\begin{pmatrix} * & * & * & * \\ 0 & * & * & * \\ 0 & 0 & * & * \\ 0 & 0 & * & * \end{pmatrix}$, and $\begin{pmatrix} * & * & * & * \\ 0 & * & * & * \\ 0 & 0 & * & * \\ 0 & 0 & 0 & * \end{pmatrix}$.

Note: $\mathbb{F}_{123}$, is the full flag manifold of lines inside 3-dimensional subspace in $\mathbb{C}^4$.

73. Consider the case G = SU(2, 2), acting on the open orbit $\mathbb{M}^+$. The holomorphic discrete series are those which are realizable as holomorphic sections of certain homogeneous vector bundles over $\mathbb{M}^+$. Let obtain the representation on L^2-holomorphic 4-forms space, where L^2, is defined with respect to the manifolds invariant inner product

$$< \omega, \eta >= \int_{\mathbb{M}^+} \omega \wedge \bar{\eta}.$$

74. Let demonstrate that the space $\mathbb{Z}^7$, is the projective space of pure spinor for SO(7, $\mathbb{C}$).

75. Let demonstrate from the exercises 72, and 74, that

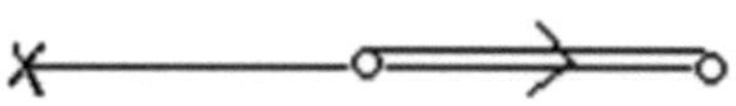
↔ CS^5

Is the quadric in $\mathbb{CP}^6$, defined by

$$Q(x) = x_0^2 + x_1^2 + \ldots + x_6^2 = 0,$$

where Q(x), is a five dimensional complex projective manifold which is a complexification of the five dimensional sphere.

76. From the two isomorphisms generates by the Penrose transform

$H^1(\mathbb{P}^+,$ 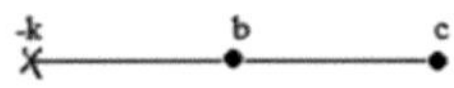

$) \cong \ker\{\mathbb{M}^+,$ 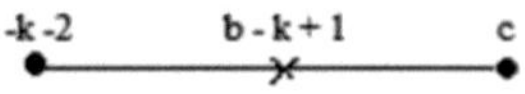

$\rightarrow$

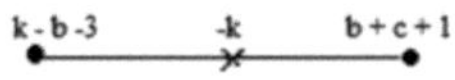

$\},$

and

$H^1(\mathbb{P}^{*-},$ 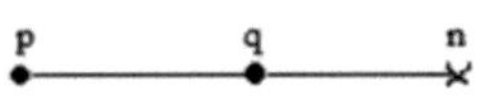

$) \cong \ker\{\mathbb{M}^{*-},$ 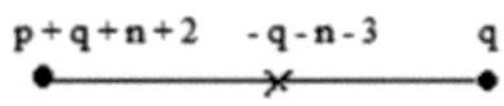

$\rightarrow$

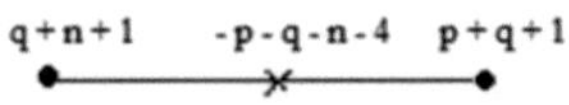

$\},$

where $k - 4 \geq b + c$, and $n \geq 0$, let deduce an isomorphism for the left-hand sides:

$\mathcal{T} : H^1(\mathbb{P}^+,$ 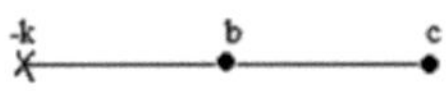

$) \cong H^1(\mathbb{P}^{*-},$ 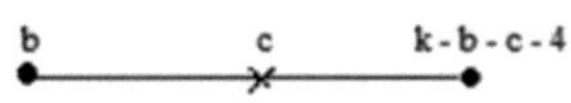

$),$

where $\mathbb{M}^{*-}$, is the component of the space-time to left-handed fields (potentials modulo gauge). $\mathbb{M}^+$, is canonically isomorphic to $\mathbb{M}^{*-}$.

"The torsion of the space-time begins in the classes of the homogeneous space G/C(T), under the twistor transform $\mathcal{T}$, which are orbits of the sided handed-fields in $\mathbb{M}$."

77. Let demonstrate that unique connect graph Γ, of an admissible set U, which can include a triplet edges is the Coxeter graph G_2,  .
78. Let demonstrate that group G, lies in the subgroup $SU(2) \subset SO(4)$, which acts trivially on V^-, and nontrivially on V^+, at the fixed point.
79. Let Z, be the twistor space of the conformally anti-self-dual manifold (X, g), and $\underline{Z}$, is the twistor space of the orbifold ($\underline{X}$, $\underline{g}$). The following vanishing theorem is the key to the structure of these two complex spaces. Let prove $H^1(Z, \mathcal{O}(-1)) = 0$.
80. For any $\lambda \in \mathfrak{h}_v$, with image $[\lambda] \in \mathfrak{h}_v/W$, we have the following:
a. The λ-twisted and λ-monodromic Hecke categories $H_{G,\lambda}$, and $H_{G,\lambda^\sim}$, are self-dual, fully dualizable Calabi-Yau algebras in St_C.

Note: The space $St_{\mathbb{C}}$, is a ∞-category of stable presentable $\mathbb{C}$-linear ∞-categories with morphisms given by continuous (colimit-preserving) functors.

b. The Hochschild homology and cohomology categories of both $H_{G,\lambda}$, and $H_{G,\lambda^{\sim}}$, are equivalent to character sheaves $Ch_{G,[\lambda]}$, with central character $[\lambda]$. The central action and trace map are given by the horocycle correspondence.

Prove them!

81. Let X, be a regular projective variety. Let C, be a hereditary Abelian category. Then for any $X \in D^b(C)$, there exists a (noncanonical) isomorphism $X \cong \oplus_{j\in\mathbb{Z}} H^j(X)[-j]$.
82. Let demonstrate that every irreducible root system is isomorphic to their dual exept B_l, and C_l, which are duals to every one of the other.
83. From the Classification theorem given in the Appendix D, table 2, let demonstrate that:
i. The number of vertex pairs in Γ, connect for at least one edge is strictly minor that n.
ii. Γ not have cycles.
84. From the before problem, let prove that Γ, not involves subgraphs of the form

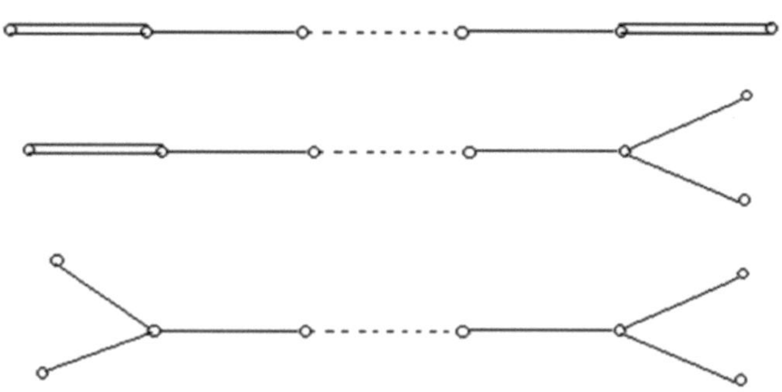

85. a). Let $\lambda \in \mathfrak{h}^*$. Let Q, be a K-orbit in X (flag variety of $\mathfrak{g}$), and τ, an irreducible K-homogeneous connection on Q, compatible with $\lambda + \rho$. Then the direct image of τ, with respect to the inclusion $Q \to X$, is the standard Harish-Chandra sheaf $\mathcal{I}(Q, \tau)$. Since τ, is holonomic, $\mathcal{I}(Q, \tau)$, is also a holonomic D_λ-module and therefore of finite length. This implies that their cohomologies $H^p(X, \mathcal{I}(Q, \tau))$, $p \in \mathbb{Z}^+$, are Harish-Chandra modules of finite length [12]. Let calculate these cohomology modules in terms of "classical" Zuckerman functors.

b). Fix $x \in Q$. Denote by $\mathfrak{b}_x$, the Borel subalgebra of $\mathfrak{g}$, corresponding to x, and by S_x, the stabilizer of x, in K. Then the geometric fiber $T_x(\tau)$, of τ, at x, is an irreducible finite-dimensional representation ω, of S_x. We can view it as an S_x-equivariant connection over the S_x-orbit $\{x\}$. Therefore, we can consider the standard Harish-Chandra sheaf $\mathcal{I}(\omega) = \mathcal{I}(\{x\}, \omega)$. It is an S_x-equivariant D_λ-module. Then let demonstrate that

$$\Gamma^{geo}_{K,S_x}(D(\mathcal{I}(\omega))) = D(\mathcal{I}(Q,\tau))[-\dim Q].$$

86. Let $\lambda \in \mathfrak{h}^*$, Q, a K-orbit in X (flag variety of $\mathfrak{g}$), and τ, an irreducible K-homogeneous connection compatible with $\lambda + \rho$. Let $x \in Q$, and let S_x, be the stabilizer of x, in K. Let ω, be the representation of S_x, in the geometric fiber $T_x(\tau)$. Then we have

$$H^p(X, \mathcal{I}(Q, \tau)) = R^{p+\dim Q}\Gamma_{K,S_x}(M(\omega)), \quad \forall \ p \in \mathbb{Z}.$$

87. Let find a relation between torsion (like field observable) and the twistor transform images on SU(2, 2)-orbits.
88. Which is the orbitalization process in non-concurrets integrals? Let investigate the non-concurrent integral concept inside the orbitalization theory.
89. What are the horo-spheres? What are the horocycles? Let explain.
90. Let (X, ω, J), be a compact Kähler manifold, and let H be a complex hypersurface in X representing twice the anticanonical class. Then the complement of H carries a nonvanishing section Θ, of $K_X^{\otimes 2}$, with poles along H. Let H, For a suitable choice of H, X \ H carries a special Lagrangian foliation whose lift to the Calabi-Yau double cover Y can be perturbed to a Z/2-invariant special Lagrangian torus fibration. Let construct an isomorphism between the foliations mentioned above and spaces in field theory inside of SO(4, $\mathbb{C}$).
91. For Penrose's quasi-local mass construction [Penrose 1982] the quasi-killing vectors are constructed out of four linearly independent solutions of the twistor equation $\omega_\alpha{}^A = (\omega_0{}^A, \ldots, \omega_3{}^A)$. They are given by $K^{AA'} = K^{\alpha\beta} = K^{(\alpha\beta)}$, is a matrix of constants and $\pi_{A'\alpha}$, are the π-parts of the $\omega_\alpha{}^A$, defined by $d\omega^A|_j = -i\pi_{A'}dx^{AA'}|_j$. The value of the Hamiltonian that generates deformations of H, with boundery value $K^{AA'}$, on j, is obtained by inserting this decomposition $K^{AA'}$, into the Witten-Nester integral

$$H(K^{AA'}) = -i\oint_j K^{\alpha\beta}\omega_\alpha^A d\pi_\beta^{A'} \wedge dx_{AA'},$$

This expression depends on $K^{AA'}$, and their decomposition into spinors. Let demonstrate that using $d(-i\pi_{A'}dx^{AA'}) = d^2\omega^A = R_B{}^A\omega^B$, we have the original definition of Penrose integral of line given in field theory.

92. Having that $\forall$ $K \subset G$, the compact component of G,

$$F_\mu(g) = \int_K a(Kg)^{\mu+\rho} dk, \forall g \in G,$$

Demonstrate that $F_\mu(g)$, is holomorphic in μ.

93. Let $H_{\sigma,\mu}$, be the ($\mathfrak{p}$, ${}^0M$)-module H_σ, with $\mathfrak{a}$, acting for $(\mu + \rho)I$, and $\mathfrak{n}$, acting for 0. Let V, be the ($\mathfrak{g}$, K)-module. Let $T \in Hom_{\mathfrak{g},K}(V, H^{\sigma,\mu})$, such that $T^\wedge(v) = T(v)(1)$, $\forall$ $v \in V$. Prove that $T^\wedge \in Hom_{\mathfrak{p},{}^0M}(V/\mathfrak{n}V, H_{\sigma,\mu})$.

94. From the before exercise 93, demonstrate that the map $T \rightarrow T^{\wedge}$, defines a bijection between $Hom_{\mathfrak{g}, K}(V, H^{\sigma, \mu})$, and $Hom_{\mathfrak{p}, 0M}(V/\mathfrak{n}V, H_{\sigma, \mu})$.
95. What is a measurable orbit?
96. The various tempered series exhaust enough of $G^{\sim}$, for a decomposition of $L_2(G)$, essentially as

$$\sum_{H \in Char(G)} \sum_{\psi \otimes e^{\vee} \in \tilde{T}} \int_{\tilde{A}} H_{\pi_{\psi, \nu, \sigma}} \otimes H^{*}_{\pi_{\psi, \nu, \sigma}} m(H : \psi : \nu : \sigma) d\sigma,$$

Prove it! Here $m(H: \psi: \nu: \sigma)d\sigma$, is the Plancherel measure on $G^{\sim}$.

97. Say what orbit type is, of the following orbits and explain which is the nature and application:
a. $D = G/H$, with D, in a generalized flag variety for $G^{\mathbb{C}}$.
b. M^{+}, of the Minkowski space $M = G_2(\mathbb{C}^4)$, that is to say, consisting of the space of 2-planes x, in $\mathbb{C}^4$, such that Φ_x, has signature (1, 1).
c. $Y = G(w)$, such that W, is a complex flag manifold where $Y \rightarrow G/P$.
d. $M_N \subset M$, with N, the nilpotent radical of an opposite parabolic subgroup.
e. The orbit defined by

$$g(x, y) = (gx, \sigma(g)y), \ \forall \ x, y \in X \times X,$$

where σ, is the involution and $g \in G$.

98. Prove that Im(*S*), is nonzero in $H^s(G/H, \mathcal{L}_{\chi})$, where *S*, is the Szëgo operator and $H^s(G/H, \mathcal{L}_{\chi})$, is an irreducible admissible representation which is a maximal globalization.
99. We assume that G, is connect. Let V, a irreducible ($\mathfrak{g}$, K)-module. Then there are $\sigma \in {}^0M^{\wedge}$, and $\mu \in (\mathfrak{a}_{\mathbb{C}})^*$, such that V, is equivalent to a submodule of a quotient module $(H^{\sigma, \mu})_K$.
100. Say what orbit type is, of the following orbits and explain which is the nature and application:
i. The orbit resulted from $D_{-}(Q) \cap \Sigma_{\lambda} = \varnothing$, and τ, satisfies the SL_2-parity condition with respect to every Q-real root in Σ.
ii. The orbit of the D_{λ}-module $I(Q, \tau)$, which is irreducible.
   i. The corresponding orbits to d, π', whose dimensions are $d(\phi)$, $d(\pi')$, where π', is an irreducible representation of a real form and ϕ, is the Langlands parameter[26] $(\phi \in P({}^{\vee}G^{\Gamma}))$.

---

[26] **Def.** A Langlands parameter is a group homomorphism

$$\phi: W_{\mathbb{R}} \rightarrow {}^{\vee}G^{\Gamma},$$

compatible with the maps into Γ, and such that $\phi(\mathbb{C}^{\chi})$, is formed of semi-simple elements:

$$W_{\mathbb{R}} = \mathbb{C}^{\chi} \times \{i, j\}, \quad jzj^{-1} = \underline{z}, \quad j^2 = -1.$$

ii. The even orbit $\underline{\mathcal{O}^{\vee}_{\vee\mathfrak{l}}}$, and a $\underline{{}^{\vee}K \cap {}^{\vee}G}$-orbit $\underline{\mathcal{O}^{\vee\mathbb{R}}_{\vee\mathfrak{l}}}$, in $\mathcal{O}^{\vee}_{\vee\mathfrak{l}} \cap ({}^{\vee}\mathfrak{s} \cap {}^{\vee}\mathfrak{l})$, such that

$$\mathcal{O}^{\vee\mathbb{R}}_{\vee\mathfrak{l}} \cap [\mathcal{O}^{\vee\mathbb{R}}_{\vee\mathfrak{l}} + ({}^{\vee}\mathfrak{u} \cap {}^{\vee}\mathfrak{s})] = \mathcal{O}^{\vee\mathbb{R}}_{\vee\mathfrak{l}} + ({}^{\vee}\mathfrak{u} \cap {}^{\vee}\mathfrak{s}).$$

101. Whih is the orbit in the integral transform of dimensions?

$$\dim F = \int_{B} \dim B^{\dim\chi(g)} \operatorname{codim} B^{0} \mu\Lambda(g).$$

Here B, is a hypersurface of the $\mathbb{C}^2$, $\chi(g)$, their index of representation induced from the corresponding vector bundle of lines and $\mu\Lambda(g)$, is a special measure through of the fractal measure followed from the vector bundle measure on lines.

102. The hypercohomology of $\mathcal{E}^{\bullet}$, of $\mathcal{E}^{\bullet}({}_{\infty}D)$, computes the cohomology of the local system $\mathcal{H}$, on M, in other words

$$\mathbb{H}^{p}(M, \mathcal{H}) \cong \mathbb{H}^{p}(\underline{M}, \mathcal{E}^{\bullet}(\log D)) \cong \mathbb{H}^{p}(\underline{M}, \mathcal{E}^{\bullet}({}_{\infty}D))\ \mathbb{H}^{p}(M, \mathcal{E}^{\bullet}).$$

Prove it!

103. Let $\mathcal{S}_{a,b}(G)$, be the space of all $f \in C^{\infty}(G)$, such that $p_{a,b,x,y,r}(f) < \infty$, for all x, y, r, endowed with the topology given by the above semi-norms. If $\gamma \in K^{\wedge}$, and if $f \in C^{\infty}(G)$, then we set

$$E_{\gamma} f(g) = d(\gamma) \int_{K} \chi_{\gamma}(k) f(k^{-1} g) dk,$$

Demonstrate $\mathcal{S}_{a,b}(G)$, is a Fréchet space.

Note: If $f \in C^{\infty}(G)$, then we set for $r \geq 0$, $x, y \in U(\mathfrak{g})$

$$p_{a,b,x,y,r}(f) = \sup_{g \in G} a(g)^{r} b(g)^{-1} |L(x)R(y)f(g)|.$$

Technical Notation

Ξ – Harish-Chandra Function

G – Lie Group of finite dimension or infinite dimension

NAK – Iwasawa decomposition of the real reductive Lie group G.

Int(G) – Inner group.

Ad(G) –Adjunct group of group G.

ad($\mathfrak{g}$) –Adjunct algebra of Lie algebra $\mathfrak{g}$.

$\mathbb{T}$ – Torus of finite dimension of a Lie subgroup of a Lie group G.

ad – Adjunct operator on Lie algebras .

[,] – Lie braket.

$\mathfrak{g}^{\alpha}$ - Proper root space of $\mathfrak{g}$, of the associated adjunct maps to the functional α.

$\prec$ – Order relation "from minor to major order".

$\succ$ – Order relation "from major to minor order".

$\mathfrak{g}$ – Lie algebra of the Lie group G.

$\mathfrak{h}$ – Cartan subalgebra if $\mathfrak{h} = \mathfrak{t} \oplus \mathfrak{a}$. A Lie subalgebra of $\mathfrak{g}$.

$Hom_{\mathfrak{g}, K}(V, W)$ – Space of ($\mathfrak{g}$, K)-invariant homomorphisms that go from module V to the module W.

$\mathfrak{m}$ – Corresponding algebra of the Lie subgroup M of the Lie Group G.

$$\mathfrak{m} = \{h \in \mathfrak{a}_F \mid Ad(g)h = h \text{ if and only if } [h, \mathfrak{g}] = 0\}.$$

$\mathfrak{n}$ – Nilpotent algebra.

H – Cartan subgroup of the Lie group.

$\mathfrak{l}$ – Levi algebra. Algebra used in the Levi decomposition of Lie algebra $\mathfrak{g}$, corresponding to the Lie group G.

G^0 – Space which arise of the identification $G^0 = {}^0(G^0)$, that is to say the space of points $\{g \in G \mid Ad(g) = I, \forall\ Ad \in End(G)\}$.

$G_{\mathbb{R}}$– Component of real points of the real reductive Lie group *G* (open subgroup of *G*).

$G_{\mathbb{C}}$ – Analytic Lie group.

F – Simple roots subspace of Δ_0.

$\underline{\mathbb{C}(\Delta)}$ –Weyl camera of simple roots space of Φ.

L – Levi group.

X(G) – Space of continuous homomorphism of G in the multiplicative group $\mathbb{R}^* = (\mathbb{R}, \bullet)/\{0\}$.

M^0 – Identity component of Lie Group M.
Space of the points: $\{m \in M \mid Ad(m) = I, \forall\ Ad \in End(M)\}$

$\otimes$ - Tensor product of modules belonging to a associative ring endowed of the tensor product to their elements.

M – Connected component of a Lie group G.

σ_α – Reflection.

$\int_G$ – Invariant integration on the group *G*.

N – Nilpotent component of the Lie group G. Also is the normal subgroup of G, when N is the normalizator of G.

N_F - Nilpotent component of the Lie group G, restricted to the simple roots subspace $F \subset \Delta_0$.

$\underline{N}_F$ – Compact nilpotent component of the Lie group G, restricted to the simple roots subspace $F \subset \Delta_0$.

$\underline{N}$ - Compact nilpotent component of the Lie group G. Also is the normal subgroup of G, when N is the normalizator of G.

P – Parabolic subgroup of Lie group G.

P_F –Parabolic subgroup restricted to the subspace F, of the simple roots of Δ_0

Cl(•) – Closure.

$^{0}(G^{0})$ – Identity component of the Lie group G^{0}.

A– Abelian Subgroup of Lie group G. Also Abelian component of the Iwasawa decomposition of the real reductive group G.

A_F – Abelian subgroup of the Lie group G, restricted to subspace F, of the simple roots of Δ_0.

$\mathfrak{m}_F$– Lie algebra of the subgroup M of the Lie group G, restricted to subspace of the simple roots of Δ_0. $\mathfrak{m}_F = \{h \in \mathfrak{a}^+ \mid Ad(g)h = h \ \forall \ g \in G\}$.

Ξ_F – Spherical function of Harish-Chandra restricted to subspace F.

$\underline{\mathfrak{n}}_F$– Compact nilpotent algebra restricted to the simple root subspace $F \subset \Delta_0$.

$T(\mathfrak{g})$ –Tensor algebra of the Lie algebra $\mathfrak{g}$.

^{0}MAN – Langlands decomposition endowed of homomorphism π, and space modulo the Hilbert space H.

$W(\mathfrak{g}, \mathfrak{a})$ – Weyl group of the homomorphisms on $\mathfrak{g}$, modulus $\mathfrak{a}$.

$L^2(G)$ – Space of integrable square representations of the group G.

dg – invariant measure on the group G (invariant under proper movements of G).

$GL(n, E)$ – General linear group on the Euclidean space of dimension n.

$A^{\mu+\rho}(k)$ – Kernel of the integral equation whose solution is a function of spherical type.

$^{*}\mathfrak{a}_F$ – Nilpotent algebra to the minimal parabolic subgroup of M_F.

$V^*[\mathfrak{n}]$ – $\mathfrak{g}$-modules belonging to the category $\mathcal{H}$.

Λ_V – Discriminant functional of the Lie algebra $^{*}\mathfrak{a}$, of the category of modules in $\mathcal{H}$.

$E(P_F, V)$ – Homomorphism space of the Osborne lemma applied to the decomposition of the algebra $U(^{*}\mathfrak{a})$.

$\Phi(P_F, A_F)$ – Space of weights of $\mathfrak{a}_F$, on M_F, of a roots system $\Phi(P, A)$.

$P(\mathfrak{a})^K$ – Polynomic Lie Algebra of all the K-invariant polynomies on the algebra $\mathfrak{g}$.

$\mathfrak{t}$ – Lie algebra corresponding to maximum torus in G.

Φ^+ – System of semisimple positive roots.

$[\mathfrak{g}, \mathfrak{g}]$ – Bilateral ideal of antisymmetric elements.

Δ – Base of simple roots of a semisimple root space.

dim – Dimension.

$(\mathfrak{a}_C)^*$ - Abelian dual algebra of Abelian algebra $\mathfrak{a}$.

$\oplus$ - Direct sum of modules or topological vector spaces.

$Z(\mathfrak{g})$ – Centre of Lie algebra $\mathfrak{g}$.

$Ind_P^G(\sigma)$ – Induced representation σ, of the group P, to the group G.

${}^0M$ – Connected component of the Lie subgroup M, of the parabolic subgroup P, of the real reductive group G.

J(V) – Jacquet module of the module V.

$\mathfrak{p}$ – Minimal compact subalgebra of the decomposition $\mathfrak{g} = \mathfrak{t} \oplus \mathfrak{p}$.

$\mathfrak{g}_C$ – Complexivity of the real reductive Lie algebra $\mathfrak{g}$.

codim – Codimension.

$Z_G(\mathfrak{m})$ – The center of the algebra $\mathfrak{m}$, explicitly defined $\{\sigma \in G \mid Ad(\sigma)x = x, \forall\ x \in \mathfrak{m}\}$.

W – Weyl group.

Ad – Endomorphism of the Lie group G. Adjunct Map on the group G.

ad – Endomorphism of the Lie algebra $\mathfrak{g}$. Adjunct map on the algebra $\mathfrak{g}$.

K – Compact subgroup of the Lie group G.

N(T) – Normalizator group of the torus T.

$\mathcal{H}$ – Category of $U^*(\mathfrak{n})$-modules.

${}^0G$ – Identity component of G. Explicitly ${}^0G = {}^0ANK$, or through continuous homomorphisms $\chi \in X(G)$, to know, ${}^0G = \{g \in G \mid \chi^2(g) = 1, \forall\ \chi \in X(G)\}$.

Conclusion

The systematical securing of spherical functions in representations theory, the obtaining of geometrical and physical properties of the space through their cycles and co-cycles, the construction of generalized functionals in complex cohomology with coefficients in a Lie algebra for the solution of the partial and ordinary differential equations in field theory, the development of intertwining integrals for the obtaining of principal representations and interrelation of induced representations (Barchini, 1992),the evaluation of integrals on groups and algebras of Lie, the calculation of topological dimensions and determination of characters of a unitary representation are only some examples of some applications that can be solved by a suitable theory of orbital integrals in the context of the theory of topological groups and their operators. Some of the said problems have given place to the development of a global harmonic analysis with the perspective to generalize the formula of Plancherel (Wallach, 1982).In the way of the study of the invariance of this formula, it was possible to have obtained a specialization of the above mentioned analysis for determination of representations located by cuspidals using the evaluation of orbital integrals on groups of Lie and their algebras. In particular the harmonic analysis that has been realized in complete form, using these methods is on the groups $SL(2, C)$, $SL(2, R)$, $SU(2, 2)$, and their compact orbits of these groups. Across a parallel study continued by the group of Harvard (Schmid, 1992) a development of induced representations has been obtained using co-adjunct orbits of a sheaf of complex holomorphic bundles, which initially was chasing to determine a method of integral transforms that were establishing classes of solutions for differential equations in field theory, using the invariance and conformability of cycles of the space-time. Nevertheless, and under a study on homogeneous spaces one manages to establish that the orbital classes are representations induced for the co-adjunct orbits determined in relative cohomology (Bulnes, 2004).Many of these induced representations could be had obtained by them through the images of integral transforms which co-cycles are G/L representations, with G, not compact and L, compact, considering certain co-adjunct orbits on au-specialization (Bulnes, 2004) that are minimal K-Types (Salamanca-Riba, 2004), (Vogan, 1992).A suitable theory of integral intertwining operators on G/L, can help to the calculation of these minimal K-Types up to certain level.

Acknowledgements

I am grateful to Demetrio Moreno-Arcega, Master of L, General Director of TESCHA, Humberto Santiago, Eng, Academic Sub-director of Tescha and Rodolfo Morales, B.L, Financing Sub-director of Tescha, for the financing and moral support to announce this mathematical research.

References

[1] N. Wallach, *Harmonic Analysis on Homogeneous Spaces*. Marcel Dekker, New York, 1972.

[2] N. Bourbaki, *Integration*. Chapitre 6, Eléments de Mathématique, Hermann Paris, 1959.

[3] F. Bulnes, *Differential Ideals: Research in Differential Equations,* Notes ofMathematical Academia and Basic Sciences, UVM, State of Mexico, 2001.

[4] C. Chevalley, *Theory of Lie Groups,* Princeton New Jersey, USA, 1946.

[5] Harish-Chandra, *Harmonic Analysis on Reductive Groups I, The Theory of the constant term,* Func. Anal, 19(1975), 104-204.

[6] S. Helgason, *Differential Geometry and Symmetric Spaces,* Academic Press, N. Y. 1962.

[7] F. Warner, *Differential Manifolds,* Springer Verlag Berlin, 1966.

[8] A. Borel et al, *Algebraic D-modules,* Academic Press. Boston, 1987.

[9] M. F. Atiyah, *Magnetic monopoles in hyperbolic space,* Vector Bundles on Algebraic Varities, Oxford University Press, 1987, pp.1-34.

[10] T. N. Bailey, and M. G. Eastwood, *Complex para-conformal manifolds their differential geometry and twistor theory,* Forum Math. 3 (1991), 61-103.

[11] M. G. Eastwood, R. Penrose, and R. O. Wells, Jr, *Cohomology and massless fields,* Commun, Math. Phys. 78 (1981), 305-351.

[12] D. Miličič, *Algebraic D-modules and representation theory of semi-simple Lie groups,* American Mathematical Society, pp.133-168.

[13] L. Kefeng, *Recent results of moduli spaces on Riemann Surfaces,* JDG Conferences, Harvard, May 2005

[14] F. Bulnes, and M. Shapiro, *General Theory of Integrals to Analysis and Geometry,* Sepi-IPN, IM-UNAM, Cladwell P. Editor, Mexico, 2007.

[15] F. Bulnes, *On the Last Progress of Cohomological Induction in the Problem of Classification of Lie Groups Representations,* Proceeding of Masterful Conferences, International Conference of Infinite Dimensional Analysis and Topology, Ivano-Frankivsk, Ukraine, 2009 (1) p21-22.

[16] F. Bulnes, *Research on Curvature of Homogeneous Spaces,* Department of Research in Mathematics and Engineering TESCHA, Government of State of Mexico, Mexico, 2010.

[17] F. Bulnes, *Integral Geometry and Complex Integral Operators Cohomology in Field Theory on Space-Time,* Internal. Procc, of 1th Internac. Congress of Applied Mathematics-*UPVT,* Government of State of Mexico, 2009 (1) p42-51.

[18] F. Bulnes, *Design of Measurement and Detection Devices of Curvature through of the Synergic Integral Operators of the Mechanics on Light Waves,* Proc. IMECE/ASME, Electronics and Photonics, Florida, USA, 2009 (5) p.91-103.

[19] F. Bulnes, *Cohomology of Moduli Spaces in Differential Operators Classification to the Field Theory,* XLII-National Congress of Mathematics of SMM, Zacatecas, Mexico, 2009.

[20] F. Bulnes, *Course of Design of Algorithms to the Master of Applied Informatics,* UCI (University of Informatics Sciences), Habana, Cuba, 2006.
[21] V. Knapp, *Harish-Chandra Modules and Penrose Transforms,* AMS, pp1-17, USA, 1992.
[22] P. Shapira, and A. Agnolo, *Radon-Penrose Transform for D-Modules,* Elesevier, Holland, 1998
[23] Grothendieck, Alexander (1960/1961). "Techniques de construction en géométrie analytique. I. Description axiomatique de l'espace de Teichmüller et de ses variantes.".*Séminaire Henri Cartan 13 no. 1, Exposés No. 7 and 8,* Paris, France.
[24] Atiyah, M. F., *Green's Functions for self-dual four-manifolds,* Adv. Math. Supp Studies 7A. 129-158, 1981.
[25] Baston, R. J., *Local cohomology, elementary states, and evaluation twistor,* Newsletter (Oxford Preprint) 22 8-13, 1986.
[26] Baston, Eastwood, M., *The Penrose transform: its interaction with representation theory,* Oxford University Press, 1989.
[27] Bateman, H., *The solution of partial differential equations by means of definite integrals,* Proc. Lond. Math. Soc. 1 (2)(1904) 451-458.
[28] Bulnes, F. J., *The($\mathfrak{g}$, K)-modules Theory,* Applied math I Sepi-National Polytechnic Institute, México, 2005.
[29] Eastwood, M., Penrose, R., and Wells.,*Cohomology and massless fields,* Commun, Math. Phys. 78 (1981), 26-30.
[30] Gelfand, *Generalized Functions,* Vol. 5. Academic Press, N. Y., 1952.
[31] Gindikin, S., *Between Integral Geometry and Twistors,* Twistors in Mathematics and Physics, Cambridge University Press. 1990.
[32] Gindikin, S., *Generalized Conformal Structures,* Twistors in Mathematics and Physics, Cambridge University Press. 1990.
[33] Helgason, S., *The Radon Transform,* Prog. Math. Vol. 5. Birkhäuser 1980.
[34] Kravchenko, V. V., Shapiro, M. V., *Integral Representations for Spatial Models of Mathematical Physics,* Addison-Wesley Longman Ltd., Pitman Res. Notes in Math. Series, v.351. London, 1996.
[35] Bulnes, F. "Conferences of Mathematics," Vol. 1., *Institute of Mathematics, UNAM, Compilation of conferences in Mathematics,* 1st ed., F. Recillas, Ed. México: 2001.
[36] Wong, H. "Dolbeault Cohomological Realization of Zuckerman Modules Associated with Finite Rank Representations," J. Funct. Anal. 129, No. 2 (1993), p428-459.
[37] Bulnes, F. "Extension of G-Modules and Generalized G-Modules," International Conference in Representation Theory of Reductive Lie Groups., IM/UNAM, Mexico, 2005.
[38] Bulnes, F. "Some Relations Between the Cohomological Induction of Vogan-Zuckerman and Langlands Classification," Faculty of Sciences, UNAM *phD. Dissertation,* ed., F. Recillas. 2004.
[39] Bulnes, F. "Conferences of Mathematics," Vol. 2., *Institute of Mathematics, UNAM, Compilation of conferences in Mathematics,* 1st ed., F. Recillas, Ed. México: 2002.
[40] Mumford, D. *Geometric invariant theory.* Springer-Verlag, Berlin Heidelberg (1965).

[41] Kashiwara, M. *Representation theory and D-modules on flag varieties.* Ast´erisque 173-174 (1989), p. 55–109.

[42] Bott, R. *Homogeneous vector bundles.* Ann. of Math. 66 (1957), 203–248.

[43] Lepowsky, J. *A generalization of the Bernstein-Gelfand-Gelfand resolution.* J. Algebra. 49 (1977), 496–511.

[44] Rocha-Caridi, A. *Splitting criteria for* g*-modules induced from parabolic and the Bernstein-Gelfand-Gelfand resolution of a finite dimensional, irreducible*$\mathfrak{g}$*-module.* Trans. A.M.S. 262 (1980), 335–366.

[45] Marastoni, C. *Grassmann duality for* D*-modules.* Ann. Sci. ´ Ecole Norm. Sup., 4e s´erie, t. 31 (1998), p. 459–491.

[46] M. Kashiwara. "Representation theory and D-modules on flag varieties,"Astérisque, (173-174):9, 55–109, 1989. Orbites unipotentes etre présentations, III.

[47] M. Kashiwara and W. Schmid, "Quasi-equivariant D-modules, equivariant derived category, and representations of reductive Lie groups, in Lie Theory and Geometry," Progr. Math. vol. 123, Birkhäuser, Boston, 1994, 457–488.

[48] W. Schmid. "Homogeneous complex manifolds and representations of semi simple Lie groups". In Representation theory and harmonic analysis on semi simple Lie groups, volume 31 of Math. Surveys Monograph. pages 223–286. Amer. Math. Soc., Providence, RI, 1989. Dissertation, University of California, Berkeley, CA, 1967.4